DRINK?

DRINK?

The New Science of Alcohol + Your Health

PROFESSOR DAVID NUTT

hachette
BOOKS

New York

Cover design by LeeAnn Falciani
Cover photograph © ViewStock/Shutterstock
Cover copyright © 2020 by Hachette Book Group, Inc.

Coauthor: Brigid Moss

Figures © 2019 Cali Mackrill at Malik & Mack

Hachette Go, an imprint of Hachette Books
Hachette Book Group
1290 Avenue of the Americas
New York, NY 10104
HachetteGo.com
Facebook.com/HachetteGo
Instagram.com/HachetteGo

First Edition: December 2020

Hachette Books is a division of Hachette Book Group, Inc.
The Hachette Go and Hachette Books name and logos
are trademarks of Hachette Book Group, Inc.

The publisher is not responsible for websites (or their
content) that are not owned by the publisher.

Library of Congress Cataloging-in-Publication Data

Names: Nutt, David J., 1951– author.
Title: Drink?: the new science of alcohol + your health / Professor David Nutt.
Description: First edition. | New York : Hachette Go Books, 2020. |
    Includes bibliographical references and index.
Identifiers: LCCN 2020029544 | ISBN 9780306923845 (paperback) |
    ISBN 9780306923838 (ebook)
Subjects: LCSH: Alcohol—Physiological effect. | Controlled drinking. |
    Alcohol—Social aspects.
Classification: LCC HV5060 .N88 2020 | DDC 615.7/828—dc23
LC record available at https://lccn.loc.gov/2020029544

ISBNs: 978-0-306-92384-5 (paperback), 978-0-306-92383-8 (ebook)

First published in Great Britain in 2020 by Yellow Kite, an
imprint of Hodder & Stoughton, a Hachette UK company

Printed in the United States of America

LSC-C

Printing 1, 2020

CONTENTS

PREFACE

THE IMPACT OF alcohol on health is profound. Up to half of all people in beds in orthopedic wards are there because of an alcohol-related injury and on weekends the emergency departments of our hospitals are filled with people who are drunk. Alcohol and medicine are inextricably entwined in the same way as money and banking.

In the UK, most people know me as the UK government chief drugs advisor who was fired in 2009 for saying government drugs policy wasn't evidence based. But I'm also a doctor, and all my professional life I, like all doctors, have been confronted with the challenge of alcohol in our patients and in our colleagues.

The real reason for my firing, I believe, was that I had the temerity to say on prime-time radio that alcohol was the most harmful drug in the UK. At that time alcohol wasn't even allowed to be considered a drug by the UK Advisory Council on the Misuse of Drugs—the ACMD—despite every scientist in the country knowing that it most certainly was a drug.

The evidence on which I based my statement was the most sophisticated and detailed analysis of drug harms ever conducted. Since then, similar studies have been carried out in Europe and Australia, each coming to the same conclusion: that alcohol is the most harmful drug in their societies too.

PREFACE

The main reason why alcohol scores at the top of the harm scale is that so many of us like to drink it. Typically, in first world Western countries, alcohol is consumed by over 80 percent of all adults. Of that 80 percent, only about one-fifth get into problems with it. But there are so many of these people that their alcohol problems have a massive impact on the rest of us, especially their families and friends. Alcohol is linked to a lot of violence both within and outside the family, road traffic accidents, loss of work, and many illnesses. In the US, drinking costs $249 billion a year, of which $28 billion are health-care costs and $23 billion from crime.[1]

Yet most of us continue to drink, and most don't get into serious difficulties with alcohol. This tells us that there are different biological and social factors that impact our relationship with alcohol. I believe that understanding these can help each of us, and our governments, make more rational and health-promoting decisions about how we all deal with alcohol. This book attempts to do this in a language every drinker can easily understand.

At a personal level, alcohol has been something I have been studying during my 40 years of medical research. For two years, in the late 1980s, I ran the inpatient research ward at the National Institute on Alcohol Abuse and Alcoholism (NIAAA) in the National Institutes of Health (NIH). Since then, back in the UK, I have continued to study the brain basis of alcohol enjoyment and problems, along with treating patients with alcohol problems.

Yet also, with one of my daughters, I own a wine bar in West London. My life encapsulates the good and bad of alcohol and pulls together my knowledge, experiences, and ideas in a way that I hope will explain why such a simple molecule as alcohol can give so much pleasure and pain at the same time.

Professor David Nutt is Professor of Neuropsychopharmacology at Imperial College London, Chair of DrugScience (drugscience .org.uk) and a former chair of the Advisory Council on the Misuse of Drugs.

INTRODUCTION

IT WAS MY first day at Cambridge, and somehow all of us first-year medical students in the same college found one another. Nine of us went to the pub around the corner.

We settled in for the evening. It was a typical 1960s city pub: small sticky tables, even stickier carpet, with a smoky atmosphere that had given a yellow veneer to the walls and ceiling. I wasn't a big drinker, but I had, I think, three pints of the local beer. Most of the others drank maybe four or five, which seemed a heck of a lot to me. At 10:30 p.m., the pub closed. Someone said they had a few bottles of wine back in their room, so we went back to the common room and hung around drinking them.

The mood was pretty loud and raucous, a bunch of green 18-year-olds having a laugh. But then, one of the guys started to cry. Not quietly but very loudly, wailing and weeping. It was an enormous outpouring of negative emotion so intense that I thought he was going to kill himself. I asked a friend who'd been at school with him, should we call an ambulance? But he said, "No, don't worry, he always does this. He won't even remember it in the morning."

He didn't remember it. But I couldn't help wondering: *How could alcohol turn someone who's a high-achieving school success, on the surface confident and likeable, into a gibbering mess?* I began to think, *what does this tell us about alcohol?* Looking back, that was the beginning of my fascination with drugs, including alcohol, and how they affect the brain, the person, and society at large.

As a doctor, I've seen firsthand all the types of havoc that alcohol causes, from liver disease and cancer to drunk-driving accidents and violence. What is clear is that you don't have to be a homeless alcoholic on the street to be harming your body and/or brain with booze. You can be a high achiever, like my university friend.

In fact, he wasn't the only one of that small group who had problems with alcohol. The school friend who reassured me he would be OK was also a heavy drinker. Very sadly, he had problems with alcohol his whole life and died of liver failure in his forties, his promising career and indeed life ruined by alcohol. True, theirs are extreme stories. But they're not particularly unusual ones.

Now, 50 years later, I want to share what I've learned about alcohol. I'm not here to preach, and I do not judge you, whether you're drinking too much or you're worried that you are; I've written this book to help you make informed and smart drinking choices. Even while I've been writing, it's made me rethink, again, how and what I drink myself. I'm not a big drinker and I know all the risks, but even so I can see how easy it is to slip into the habit of overindulging.

My position has always been: Alcohol is a drug and one that shouldn't be taken lightly. This may sound odd in our culture, where having a drink and indeed getting drunk is broadly acceptable. But it's a position most doctors, scientists, and addiction experts take too, given the enormous amount of damage alcohol does to people's health and to society.[1]

The World Health Organization's *Global Status Report on Alcohol and Health 2018* said that in 2016 more than 3 million people died as a result of harmful use of alcohol. A recent editorial in the *British Medical Journal* (*BMJ*) sums up how: "Like tobacco, alcohol kills some users slowly through the diseases it causes. Unlike tobacco, alcohol also kills quickly, through injury and poisoning."[2] Alcohol causes more than 5 percent of the global disease burden. It's estimated that alcohol contributes to more than 80,000 premature deaths in the US per year. The big killers include cardiovascular disease, which leads to heart attacks and stroke. Then there are liver disease and various cancers, as well as accidents, especially on the road, and deaths from suicide.

I'm not trying to scare you, I just want you to know the facts so you can make your own decisions. Thirty percent of Americans don't drink at all and 30 percent consume on average less than one drink (14 grams alcohol) per week; at the other extreme, the top 10 percent of American drinkers consume an average of 74 drinks per week, that is, over 10 drinks per day![3]

In the UK, two-thirds of people who drink more than eight US standard drinks per week say they'd find cutting down on drinking harder to do than other lifestyle changes like exercising more, reducing smoking, or improving their diet. If you have picked up this book, I'd imagine you might be one of them. Perhaps, in the past, you've felt relieved when you've seen a newspaper article about "moderate" drinking, particularly red wine, being good for your health. But the truth is, as you'll see in Chapter 2, no level of alcohol consumption is safe.

You can dress alcohol up as an Aperol spritz or a glass of Pol Roger, or dress it down as a 40-ounce bottle of malt liquor, but either way, it's a molecule that taps into our brain chemistry to produce a whole range of effects. Including, in some people, addiction. Until you appreciate that, you run the risk of underestimating its power.

The shift in the way we consume alcohol in the last 50 years, the length of my career, is a remarkable example of how marketing has altered our perception. Alcohol used to be a special purchase—you went to a bar to buy it. Now in many US states it's widely available in supermarkets, and many people see it as part of their standard weekly grocery shopping.

It's not until you think about it that you realize how alcohol is entrenched in every part of our lives. We drink for social bonding. We drink together to clinch business deals and come to agreements. We drink to celebrate the birth of a child, to commiserate with one another when someone dies.

When one of my nephews turned 18, I went to buy him a birthday card. I counted 23 "Happy 18th!" cards before I found one that didn't focus on alcohol. What kind of message is that for a young person? Luckily, it seems as if that age group—or some of them—are drinking less.[4] Because it's absurd that coming of age should be all about alcohol. Still, those cards show that reaching the age of majority when you are allowed to drink is seen as one of the great transition points in a person's life.

Alcohol has glamour and history, and our art and culture are steeped in it. It's thought to be nearly as old as human society. There's a theory that the roots of agriculture weren't in the search for food but in the cultivation of crops to make alcohol.

INTRODUCTION

The Romans had their Bacchanalia, the Greeks their symposia, both fueled by wine. We have our bachelor and bachelorette parties, wine tastings and brewery tours, office Christmas parties and dinner parties.

In the thirteenth and fourteenth centuries, the Catholic Church was the main source of alcohol as monks held a lot of the cultural knowledge of wine- and beer-making. The oldest continuously produced beer goes back nearly a thousand years and is from a monastery in Germany.[5] But now, one company produces a third of the beer in the world.[6]

Drinking—and drinking heavily—is in our cultural references, from the paintings of Toulouse-Lautrec to James Bond. Toulouse-Lautrec's favorite drink was reportedly absinthe, the Parisian drink of the moment, mixed with brandy. And if I drank as many martinis as 007, I'd be shaken too. He should be called the man with the golden liver!

Sure, not all cultural references are positive. In the novel *L'Assommoir* by Émile Zola, a woman desperately tries to hold her family together while her husband becomes an alcoholic, but she ends up dying in penury as one herself. In the critically acclaimed film *Judy*, Renée Zellweger brings to life the sadness of the final alcoholic years of Judy Garland.

The only difference between alcohol and any other drugs, I've always argued, is that alcohol is legal. That position has gotten me into trouble—neither the government nor the alcohol industry want alcohol to be treated as a drug, for obvious reasons—but I am a scientist, and when you look at how it affects the neurotransmitters of the brain (see Chapter 1), it's clear it's pretty close in its brain effects to quite a few other drugs; although it's more promiscuous than any other. No other drug both enhances GABA, serotonin, and dopamine and at the same time blocks glutamate and noradrenaline. Alcohol truly produces a cocktail of neurotransmitter effects.

In three papers I cowrote between 2010 and 2019,[7] we showed that alcohol is the most harmful drug in the UK, Europe, Australia, and so probably in all the Western world. We did a decision analysis to rank legal and illegal substances, including heroin, cocaine, crack, and crystal meth, as well as alcohol and tobacco. Alcohol scored the highest not because of the harm it does to the individual but mainly because of the vast harm it does to others—passive drinking, if you like.

In my first ever proper job as a psychiatry trainee, working at the emergency department at Guy's Hospital in southeast London, one of my very first patients arrived in a terrible state, borderline psychotic. She'd come straight from her wedding reception, where, in a drunken brawl, her brothers had beaten up her new husband. Sadly, that is a story repeated in various forms in emergency departments across the UK on Friday and Saturday nights.

It is true that the huge amount of harm from alcohol is largely driven by the number of people who drink. If half of adults took crystal meth, for example, society would be in a much worse state. But on a personal level, there are very many harms of alcohol, and these are determined by how much you drink. They may not be obvious, like the link to breast cancer that has only recently become clear, or the 10 to 20 years that drinking takes to damage your liver.

You may be hoping none of this will apply to you. Most of us love alcohol's short-term effects—the fun, the sociability, the relaxation—and hope the long-term health damage won't happen to us.

STANDARD DRINKS IN THE USA

In the US, a standard drink—the measurement used to count alcohol consumption—is equal to 14.0 grams (0.6 ounces) of pure alcohol. Generally, this amount of pure alcohol is found in

12 ounces of beer (5 percent alcohol content)

8 ounces of malt liquor (7 percent alcohol content)

5 ounces of wine (12 percent alcohol content)

1.5 ounces or a "shot" of 80-proof (40 percent alcohol content) distilled spirits or liquor (e.g., gin, rum, vodka, whiskey)

But if you are drinking above (or even below) the recommended level of **no more than one standard drink a day for women and two for men,** you need to know the risk to your health. I hope this book will make this clear. The alcohol industry knows alcohol is a toxic substance. If it were discovered today, it would be illegal as a food product. The safe limit of alcohol, if you applied food standards criteria, would be one glass of wine a year.[8] Would you take a new drug if you were told it would increase your risk of cancer, dementia, heart disease, or that it shortened your life? You wouldn't touch it. But alcohol has a special place in our culture.

What I would like to help you do is think about your drinking. To know what the risks are if you go above the recommended limits. To know that even if you can't stick to the limits all the time, you should not just give up and ignore them completely. To learn how to monitor your drinking. Perhaps, sometimes, to choose not to drink.

What, ideally, I don't want you to do anymore is to throw alcohol down your throat without thinking. When you've read this book, you may decide to join the growing number of people who decide not to drink. Or you may decide you are happy drinking exactly as much as you are now, but you will have made that decision in possession of all the facts.

Because what drinking alcohol needs to be, above all, is a conscious act. You should treat it as more special than eating, more like—I would argue—it used to be treated in the past. Make drinking a positive, active pleasure rather than a reflex and habit, or something you've always done, or self-medication for stress or anxiety.

My view is, try to reduce how much you are drinking as much as you can toward the recommended limit while maximizing your fun and pleasure. And have at least two days off a week. I hope I'll give you ideas for how to do that.

I do not blame you if you drink too much. Not only does society introduce us to drinking at an early age and tell us it's the most fun we can have, but as you'll find out (see Chapter 7), it's very easy to become psychologically and/or physically addicted to alcohol.

I hope this book will show you how to drink for living rather than to live—or die—for drinking. I hope it will help you start to take back your agency over how much you drink, to be able to decide how much is right for you—then drink exactly that. Cheers to that!

HOW DRINKING AFFECTS YOUR BODY AND BRAIN

FROM FIRST SIP TO THE MORNING AFTER: ALCOHOL'S JOURNEY

LET'S TAKE A short tour through your brain and body on alcohol. Your first drink starts with a single mouthful, to bastardize a saying by Mao Zedong.

In fact, you would be surprised how difficult it is to get people to drink alcohol for the first time. Pure alcohol tastes repellent, and nobody—except perhaps an alcoholic—would drink it. Even in the various forms in which we buy it, it's often bitter and tart, if not downright unpleasant, and needs a mixer. Some of the common flavors that alcohol comes in—wine-flavored, hops-flavored—are quite strange and so also have to be acquired.

One of the most effective ways to make alcohol more appealing to people is to add sweeteners. This holds true for rats too. When scientists use rats to study the effects of alcohol, it's standard to sweeten it to ensure the rats drink it—a rat cooler, if you like. In fact, I'd say the main reason super-sweet (and often brightly colored) coolers like Smirnoff Ice are on sale is to appeal to entry-level consumers who don't yet have the taste for alcohol, aka teenagers.

You may also be surprised to hear that we quickly acquire that taste. Very soon after the first mouthfuls, our brain learns that a few minutes after the difficult taste come good feelings of being warm, relaxed, maybe a little more sociable. And once those good feelings become associated with the taste, the smell, and the setting, routine, and/or ritual, we then start to like the taste. I've often heard people say things along these lines: "I love the taste of 1984 Château Latour/Whispering Angel/Pol Roger." But I say to them: "If you gave that to your

IS MORE EXPENSIVE BOOZE BETTER FOR YOUR BODY?

You can't buy your way out of harming your body. There may be an argument that red wine and beer contain some health-protective factors (see Chapter 2 for more), but they don't contain enough to make up for the harms of ethanol, alcohol's most toxic chemical. If you get less of a hangover when you upgrade your drinks, it may be that you are savoring it more and so drinking less; which is no bad thing. But sorry, the short answer is no.

child, they would spit it out. You have acquired the love of that taste. And what has given you that love of the taste is the effect of the alcohol. And, of course, the knowledge that it's really expensive."

This learning effect holds true in rats too: once a rat has been made to drink enough alcohol, it will continue to drink it, even if it's not flavored.

Another sensation many people learn to like is the hot feeling you get as alcohol goes down your throat, which will be familiar if you drink spirits and especially if you've ever done shots. It can become appealing because your brain knows you're about to get the pleasurable hit from the alcohol. As spirits are a dilute form of alcohol, that hot feeling is not nearly as intense as if you were to put neat alcohol on your tongue or a cut. That would burn—and really hurt.

These two elements—the flavors and the mouthfeel as well as the look of your favorite drink and the place and time you usually drink—prompt your brain to get ready to experience the effects of alcohol.

As the liquid flows into your stomach, it begins to be absorbed through the walls of the stomach, then via the small intestine. The alcohol goes via the bloodstream into the liver, where it starts to be broken down, the main by-product being acetaldehyde. Then this alcohol and acetaldehyde mixture travels through the bloodstream and into the heart and also crosses the blood–brain barrier and enters your brain.

Alcohol is such a small molecule that absorption happens pretty quickly; within five or ten minutes of your first mouthful, you will start to feel the first effects in your body. You may begin to feel warmer, a little flushed, as alcohol allows the blood vessels of your skin to expand, which is called vasodilation.

INSIDE YOUR BRAIN

When the alcohol mixture hits the brain, it starts to do the things that make us like it. I find it helpful to describe the brain as working like an electrochemical machine. This machine is made up of a web of around 200 billion neurons. All our thoughts and processes are mediated by messages between the neurons. Your machine's outputs—being awake, asleep, storing memories, swallowing, and so on—are the results of these millions of messages, flying around the brain web.

Each message travels along the neuron via electricity, but the connection that bridges the gap between neurons—called the synapse—is chemical. The chemicals that bridge those gaps are called neurotransmitters.

What's important to know for the story of how alcohol works is that it—like other drugs—works at the level of this chemical connection.

There are around 80 different types of neurotransmitters (chemical messengers). And there are even more types of the receptors that they slot into. Each receptor is triggered by a different neurotransmitter.

The two neurotransmitters that are the most common and the most powerful, because they are effectively the on–off switch of the brain, are gamma-aminobutyric acid (GABA) and glutamate. In essence these two neurotransmitters are the core of the brain. They do all the basic work such as sleeping, laying down memories, and thinking.

Glutamate turns on the brain and GABA turns it off. When glutamate is released and goes across the synapse, it turns on the next neuron, which makes the brain more active. GABA does the opposite.

The two work hand in hand, like yin and yang. You need GABA to control glutamate because if you have too much glutamate your system goes into overload; this leads to getting very anxious, or even to a seizure, and potentially to brain damage.

Your brain has evolved so that every time glutamate is released, GABA is released too. It's a beautifully balanced system.

There's another type of neurotransmitter, called a neuromodulator because it modifies the brain's response rather than affects it directly. So, if you have a car accident, the memory of it will be laid down by glutamate—for example, that you were driving at 60 miles an hour, that you were on the highway—but the emotion will be encoded by a neuromodulator, noradrenaline. This adds extra information to the memory, but it's not the core component.

Neuromodulators work like a backup to GABA and glutamate because most outputs of the brain use more than one neurotransmitter.

To explain with an example, we think there are at least four neurotransmitters that keep us awake: acetylcholine, orexin, histamine, noradrenaline—and at least four that put us to sleep: GABA, adenosine, serotonin, endocannabinoids. This means that if one fails, the others can back it up. So someone with narcolepsy, for example, has a deficiency in one neuromodulator, orexin. They're not good at staying awake because the backups aren't always powerful enough.

The names of some neuromodulators will be very familiar

to you. As well as noradrenaline, serotonin and dopamine are involved in many processes, particularly emotional ones. Then there are the endorphins, which are actually a different type of molecule called a peptide, little pieces of protein.

All the various drugs and mind-altering substances that human beings take work on different combinations of neurotransmitter and neuromodulator systems. Alcohol is one of the most promiscuous of drugs, in that it affects a lot of different types of receptors and hence the majority, if not all, of the neurons. Which is why it can give us so many different kinds of effects and experiences.

YOUR BRAIN ON ALCOHOL

As you can see, the brain is an incredibly evolved and finely balanced machine. And then you add in alcohol and that balance dissolves like a sugar cube in a hot cup of tea—or rather, a hot toddy.

The first thing alcohol does is to turn on the calming GABA system, so you start to feel relaxed. Valium works on the same system. This is why we drink. And it's especially why a lot of us drink at parties.

Most of us have some level of social anxiety, and alcohol removes our fear and inhibitions. Alcohol's calming effect is also the reason why, as soon as the seatbelt sign is turned off on an airplane, the beverage cart is wheeled down the aisle. It's because many people are anxious about flying.

However, if you turn the GABA system on too much, it can switch off parts of the brain you don't want switched off, for example, your judgment or even your consciousness. And if you drink a great deal of alcohol and the GABA system is maximally potentiated, it turns off your brain, in the same way as an anesthetic, so you stop breathing. That's one way you can die from alcohol. This is what we call alcohol poisoning.

Going back to your judgment, do you ever intend to drink a couple of drinks but then lose control and end up bingeing? Why is that? One of the main reasons is that the part of the brain that tells you to stay in control—the frontal cortex—is the first part that's switched off by alcohol.

In fact, the parts of the brain that are about retaining judgment and control are very sensitive to alcohol. You may find you lose judgment over other things too: your attractiveness, your ability to dance or to chat someone up.

Drink more, and as you go over the drunk-driving limit—which is 80 milligrams of alcohol per 100 milliliters of blood (a blood alcohol content [BAC] of 0.08 percent)—you begin to get the double whammy: as well as stimulating GABA, the alcohol starts to block your glutamate receptors. And remember, glutamate is the neurotransmitter that keeps you awake. As your level gets higher, you're starting to become properly drunk. If you reach the point of 0.15 percent BAC, you'll also start to lose the capacity to lay down memory. This is a blackout.

As your blood alcohol concentration rises, it affects new and different neuromodulators too. And each one will have its own specific influences in different parts of the brain.

There are three that are the most important. First, a rising blood alcohol level increases the effects of serotonin, a mood enhancer that also makes you more empathetic. Its pro-serotonin effect is also what makes other people seem more attractive—so-called "beer goggles." In that way, it has a similar effect to MDMA, or ecstasy.

It's the stimulation of a different serotonin receptor, one in the nerves of the stomach, that makes you sick.

We talk about throwing up from alcohol pretty casually, but, in fact, vomiting is crucial, as it stops you from dying. This survival mechanism is one of the reasons alcohol has survived in our culture for thousands of years. Vomiting gets rid of enough alcohol so you stay alive.

Second, drinking releases dopamine, which is involved in drive, motivation, and energy. This is a factor in alcohol's stimulant effect, which makes you feel exhilarated, more active, and gives you feelings of energy and enthusiasm. Dopamine makes you louder—this is an effect people get from cocaine too.

Dopamine is one of the transmitters that also lays down behavioral patterns, so it's important in addiction. Dopamine release may be the reason people get locked into habits that start off being fun, or at least not damaging, but then become so, for example, hair-twisting and nail-biting. Dopamine is also why you get into stupid arguments about irrelevant things when you're drunk, but can't stop yourself.

Third, the high you get from alcohol comes from endorphins. These are your body's natural opioids, the brain's natural pain-reducing system, also the source of the runner's high. This reward system gives you a chilled sort of pleasure and, in some people, may also be a key factor in their addiction. Several studies have shown that when the effects of endorphins are blocked with the anti-drinking medication nalmefene, some addicts are able to stop drinking.[1] Using sophisticated brain imaging, we have been able to see that this effect is related to the interactions between endorphins and dopamine in the brain.

It's this cornucopia of effects that gives alcohol such a wide appeal. And as we are all different, it will slot into your personal brain chemistry in a different way to that of your friend. Perhaps you drink to reduce anxiety? Or after work as a valve to release stress? Or to get motivated to go out? Or for liquid courage to go fighting? A large part of alcohol's appeal is that, for many people, it fills in the gaps in your personality, making you the person you want to be.

THE MIND-BENDERS IN YOUR BOOZE

What you drink is as full of psychoactive substances—chemicals that change your brain chemistry—as any other drug. By far the most abundant alcohol in the bottles on liquor store shelves is ethanol (chemical formula C_2H_5OH). During fermentation, the glucose in the raw ingredients breaks down mostly into ethanol, but it will always make other types of alcohol too, which differ in amounts and chemical structure in different drinks. Even what is sold to us as the purest alcohol, for example, vodka, contains a cocktail of various alcohols. The only pure alcohol is ethanol that's been synthetically produced.

So, for example, when whiskey has been analyzed, it has been found to contain roughly 400 different alcohols. This is because the longer a whiskey is stored—or "aged"—the more some of the alcohols will join together to form more complex alcohols called congeners. These exist in all alcoholic drinks (although whiskey contains the most) and are thought to work similarly in the brain to simple ethyl-alcohol (ethanol) but may perhaps be even more intoxicating. Each whiskey will have its own combination of alcohols and congeners, and it's this mixture that creates all the various nuances of its "nose" and so its flavor.

Interestingly, under the new UK Psychoactive Substances Act 2016, every recreational psychoactive substance was banned except ethanol, caffeine, nicotine, and tobacco. At the time of the Act, I did say that this was absurd. In effect it makes all alcohol illegal because there is not a single alcohol you can buy that doesn't contain congeners. Beer has around 150; wine has 200. We don't drink pure ethanol.

As well as their own blend of hundreds of different alcohols, wine, beer, and cider also contain aromatic plant compounds called terpenes, which come from grapes, hops, or apples. You may have heard of terpenes as they give cannabis its distinctive smell too. It used to be assumed that terpenes contributed purely to flavor, but it's now thought possible—though it isn't yet well studied—that they are also psychoactive.

SO YOU KEEP DRINKING . . .

Keep drinking and, by this point, you are likely to be slurring. Your memory is shot due to the glutamate effect, so you're repeating yourself, repeating yourself.

You may also be finding a lot of things funny, including the things you're repeating, which may be due to your serotonin and GABA systems. And alcohol is dampening down the centers of the brain that control coordination, hence the UK slang "legless."

With your dopamine up and your self-control down, you may be beginning to get argumentative. You may also do something stupid—smoking when you've given up, drunk driving, shopping-cart racing—because your judgment is impaired.

There was a really sad case of a 16-year-old girl, Natalie Dursley, who was picked up by an ambulance after collapsing at a nightclub. She was so confused, she opened the doors of the vehicle on the highway, fell out, and died. This is a perfect example of the complete loss of insight that can happen when you are drunk. At root, you've seriously disrupted the efficiency of your brain—an effect I like to compare to having a virus in your computer.

Keep on drinking, and you're moving toward anesthesia, a system shutdown. In fact, medical anesthetics target the GABA and glutamate systems too. An anesthetic you'd have for a minor op would switch on GABA to put you to sleep. And for a major one, it'd switch off glutamate, which keeps you awake and alive. That is why you need to be ventilated, as you can no longer breathe on your own.

Before modern anesthetics existed, if a sailor had to have an injured limb removed, he'd be made legless first. There's some good reason for this: alcohol does dampen down the pain—as well as the memory.

It's the fact that alcohol affects both GABA and glutamate—the double whammy—that makes it dangerous. GABA and glutamate regulate not only being awake but also being alive. This is why people do die of alcohol poisoning. If you drink enough, you can stop breathing. Below is an at-a-glance guide to how drunkenness progresses.

WHAT HAPPENS AS YOU DRINK MORE?
THE STAGES OF DRUNKENNESS

As your blood alcohol content rises, it affects the way you feel and behave, as well as the different neurotransmitters in your body.

blood alcohol concentration / you will feel / neurotransmitter(s)

- 0.02 percent / relaxed, slightly altered mood, slightly warmer / GABA + endorphins
- 0.05 percent / less inhibition, louder speech, more gesturing, reduced coordination and eye focus / GABA + serotonin
- 0.08 percent / drunk-driving limit. Loss of coordination, balance, speech, hearing, and reaction times / GABA + endorphins
- 0.1 percent / slurred speech. Reduced reaction times and physical control / GABA + dopamine
- 0.15 percent / euphoria, coordination reduced so much that you may fall over. Walking and talking become hard. Possibly vomiting / Serotonin + dopamine + GABA
- 0.25 percent / confusion, stupor, disorientation. Loss of pain response. Nausea and vomiting may begin. Hard to stand and walk without help. Blackouts begin / glutamate block + GABA
- 0.3–0.4 percent / possibility of falling unconscious / glutamate block + GABA
- 0.4 percent / stupor / glutamate block + GABA
- 0.5 percent / coma / glutamate block + GABA
- 0.6 percent / breathing stops and death / glutamate block + GABA[2]

CONTEXT IS EVERYTHING

Different people do have very different experiences under the influence, which comes from the setting and expectations more so than from any difference in the alcohol itself. The buzz from a warm can of gin and tonic on a packed commuter train will feel completely different from the same measure in a martini in a chic hotel bar, for example. And pregaming before clubbing is a world away from a glass of sherry with your parents at Christmas. People might drink at bedtime to relax or they might (though it's best not to) drink before a business meeting for confidence.

Buckfast Tonic Wine is interesting as a social experiment (albeit inadvertent) and a great illustration of how alcohol's effects interact with personal intention. Originally made by the monks of Buckfast Abbey in England as a pick-me-up, "Buckie" is a strong, fortified wine—15 percent alcohol—with lots of added caffeine. One bottle contains the equivalent of nearly five double espressos.

It was first sold as a tonic for people who needed energy, as the caffeine overcomes the sedative effects of the alcohol. You might know that Buckie has become the cult drink of soccer fans in Scotland (where it is also known as "wreck the hoose juice"); these fans down a bottle (or maybe more) before going out for a fight. In a Scottish survey of young offenders, 43 percent who'd drunk before their offense said their drink of choice was Buckfast.[3]

It's the same wine as when it was marketed as a tonic but is now used for a completely different purpose. In 2014, the Scottish government debated whether Buckie should no longer be sold in glass bottles, as there were reports of them being used as weapons, according to the *Scotsman* newspaper.[4] Although this law was not passed and Buckie still comes in a wine bottle, it is now—perhaps to reduce the risk of broken bottle violence—available in cans as well.

Of course, violence and alcohol are well-established bedfellows. It's one reason most sporting venues and festivals no longer use glasses (the other is that drunk people are so clumsy).

TIMING IS ALL

The time of day you drink is also a factor. It makes sense that the sleepier you are, the more likely it is that alcohol will put you to sleep. That's the reason people mix stimulants and alcohol as in Buckie; stimulants to keep you awake, sedatives to take away the anxiety. In fact, "uppers and downers" is the most popular combination in the history of drug taking.

Another example of this is cocaine and alcohol. Back in the 1890s, when cocaine was legal, a wine called Mariani from Italy contained both. And it was endorsed by the Pope, no less. And you don't need to have seen *Scarface* to know that people mix cocaine and alcohol at parties. The reason why people might do this is in order to be able to drink more and for longer. Interestingly, when in the 1990s the Icelandic government passed a law to allow 24-hour drinking, there was subsequently an increase in amphetamine use.[5]

One of the issues with cocaine and alcohol is that they work together in the body to produce a new chemical, called coca-ethylene (CE). CE is a longer-acting form of cocaine that hangs around in the body for hours rather than minutes. And this makes it more toxic to the heart. That's why there's such a strong association between taking cocaine, drinking, and heart attacks.

The upper–downer effect accounts for the popularity of Red Bull and vodka too; it's a mix that's been shown in animal studies to have brain-altering effects.[6] You might consider a red bull vodka terribly unsophisticated and prefer an espresso martini, but in reality, what you're doing to your brain is not so different.

Alcohol with tobacco is probably the most common combination of all. A lot of people find that once they've gotten into the habit of having a cigarette with a drink, it's hard to have one without the other. This may be due to the fact that smoking accentuates the impact of alcohol on dopamine. It may also be due to the fact that alcohol disinhibits you, so after a drink you lose the will not to smoke.

There is no doubt, either, that getting drunk faster—and so raising your blood alcohol levels faster—will mean you overcome your inhibitions more quickly. Drink fast, and you may find your insight has gone and you are overwhelmed by alcohol before you even realize you are drunk. The strength of the alcohol you're drinking also matters in this respect; the faster you are able to get drunk, the more quickly bad things start to happen.

This is why the trend for drinks becoming stronger in the past 50 years or so is not a positive one. In the 1960s and '70s, when I was a student, most lagers and ales in pubs were 3 to 4 percent. Then along came Stella at 5 percent. Now pub beers are routinely 4 to 5 percent. Wine's strength has also generally gone up, from 11 or 12 percent to 13 or 14.

Some drinks have been shown to make your blood alcohol levels rise faster for other reasons too: champagne more than wine, for example. In fact, most people get drunk faster on fizzy drinks.

THE HIDDEN DANGERS OF TOLERANCE

The toxicity of alcohol will vary according to your genetics but also your past drinking experience. This is your tolerance, your body and brain's ability to prepare for and withstand alcohol. Tolerance builds up very fast because your brain quickly learns to expect alcohol. At the start of a vacation you might feel drunk on, say, a glass or two of wine. By the end of the week, it might take a bottle to feel the same way.

A high tolerance always has its dangers. I once went to stay with a friend, and when we sat down to watch rugby he offered me a drink. When I said yes, he poured me a full wine glass of Scotch, which I sipped. He poured himself the same measure and, while I sipped mine, kept finishing it and pouring himself more. By the end of the match, he'd drunk a full bottle of whiskey (minus my one glass!). His tolerance was sky-high.

The sad story of Amy Winehouse illustrates another danger of tolerance: the consequences of losing it. Coming out of six weeks in treatment, she went back to drinking the amount she had done before rehab. But because she'd lost her tolerance, this amount of alcohol had become enough to poison her.

If you think about it, tolerance is really dependency—or at

least the beginnings of it. If you look back and realize your drinking has been steadily increasing, take a look at Chapter 7, where I explain more about how tolerance works.

WHY DO I GET DRUNK BUT MY FRIEND DOESN'T?

There could be a few reasons—and there are likely to be more than one. Not least, there's tolerance: your friend might be used to drinking a lot more than you, so their brain will be prepared to combat the effects of alcohol. If your friend is a different sex from you, that will make a difference. If a man and a woman drink the same, the woman's blood alcohol level will be higher. It's not only because women tend to be smaller but also that, proportionally, women's bodies have a higher percentage of fat and a lower percentage of water, and alcohol is diluted in the body's water content.

How much your friend has eaten and what they've eaten will make a difference too, as food delays alcohol uptake from the stomach. This is the reason why some people swear by drinking a pint of milk before going out on a heavy night. And it's not just perceived wisdom. True, the milk won't have a huge effect, but the fluid will fill them up and hydrate them so they're less thirsty. The fat, as with other fat-containing foods, delays stomach-emptying. As only 20 percent of alcohol is absorbed in the stomach, this slows alcohol uptake into the blood.[7,8]

Then there are genetic factors. Your friend's genetics may mean they metabolize alcohol faster than you. Some people have trouble metabolizing alcohol, in particular the stage after the body turns it into acetaldehyde. This inability is most common in people of Japanese, Chinese, and Korean descent.[9] The result is, they get the bad effects fast, in particular looking flushed and feeling very drunk quickly.

Some people, particularly the sons of male alcoholics, seem to inherit brains that are subsensitive to alcohol. This means that they can tolerate more than their peers from the very first time they drink. It's a kind of pre-tolerance that seems good to start with but over time can lead them to drink more and so be at a greater risk of alcohol damage.

THE AFTERMATH . . .

WHY YOU EAT MORE WHEN DRUNK

Do you often end the night going for a slice of pizza or a burger or your drunk food of choice? It's not well understood why we get so hungry when we've been drinking.

One reason might be that alcohol changes the production of sugar in the liver so you become hypoglycemic; in other words, you get low blood sugar so you feel hungry. Or it may be the case that eating, like smoking, is a way of boosting your reward-pleasure experience. Or if you are dieting or restricting food in some way, it could be that alcohol's inhibition-reducing effect on GABA reduces your resolve not to eat. There's probably a similar reason why drunk people are more likely to have sex too.

WHY YOU WAKE UP TIRED AND WIRED

As it's a sedative, alcohol puts you into very deep sleep. This explains why, after they fall asleep drunk, people do odd things, such as get up and pee in the closet. They are in such a deep sleep that their conscious brain stays asleep even while the part of their brain that's telling them to pee is awake. And as night terrors—intense nightmare-like experiences where you can still move—happen in this deep sleep state too, if you're prone to them, alcohol can provoke them (more on this in Chapter 6).

The sedative effect works for around four hours. At this point, you may wake up with a headache and perhaps with lights flashing in your eyes, feeling alert and unhappy, and find it hard to get back to sleep. If you don't wake up, you're likely to sleep badly or fitfully.

This is happening because your brain knows being intoxicated is potentially dangerous—so it doesn't want to be intoxicated. As soon as you started to drink the previous night, your brain started to change. Because alcohol blocks glutamate receptors, your brain up-regulated the glutamate system to compensate for this, by increasing the quantity of glutamate receptors.

Then, as your blood alcohol level goes down during the night, you're left with too many receptors and so too much glutamate activity. And that is why you are too awake and alert, and why the world seems too bright, too noisy, too much.

WELCOME TO YOUR HANGOVER

Good morning! Or likely not. People joke about hangovers—
"What's the best thing for a hangover? Drinking heavily the
night before." There's even a multimillion-dollar film fran-
chise based on the confusion, sickness, and regret we experience the morning after. But what is happening is more serious
than that. It's withdrawal. Yes, the symptoms are less extensive, unpleasant, and life-threatening than an alcoholic will
go through. But however you dress it up, your brain and body
are withdrawing from alcohol.

The definition of an alcohol hangover is the experience of
various unpleasant physiological and psychological effects
that follow the medium-to-high consumption of alcohol.
Typically, it comes on around ten hours after your blood alcohol peaks—but this varies according to sex, weight, and your
genetic disposition too. Some people don't get them at all:
between 3 percent[10] and 23 percent[11] of the population are
reported to be hangover resistant. Not surprisingly, these
people may be more likely to become heavy drinkers, as they
don't experience the deterrent effect of hangovers.

As you might have experienced, hangovers can last from a
few hours to over 24. And as alcohol affects so many systems in
your body and brain, a hangover is equally complex, hence the
myriad possible ways to suffer. There are over 47 possible symptoms (see the table on page 31), and you're likely to get your
own personal selection of them; they range from sleep disturbance and dehydration to anxiety, tension, and negative
emotional state to impairment of attention, memory, and
psychomotor skills. Symptoms will depend on how much you've

drunk but also on what you've drunk. Cask-aged and matured alcohols have a lot of congeners (which we looked at earlier), and it is thought that these may create worse hangovers.

The upshot? Even though you're no longer drunk, you probably won't be much use at work and you almost definitely shouldn't be driving. The US Centers for Disease Control and Prevention estimated that alcohol hangovers cost the US economy approximately $180 billion in 2010.[12]

WHAT'S HAPPENING IN YOUR BRAIN AND BODY

There are gaps in the research about what is going on in your body during a hangover. What we do know is that the hangover state is a multifactorial event caused by a variety of biochemical and neurochemical changes as well as your personal genetic makeup.

1) YOU'VE POISONED YOURSELF One of the ways alcohol is metabolized is by the enzyme alcohol dehydrogenase (ADH). As this enzyme breaks down ethanol, it forms acetaldehyde, a poison and carcinogen. This is relatively quickly turned into acetate, then finally into carbon dioxide and water. However, some people have genetic variants of the relevant enzymes that make the breakdown faster or slower. What's known from studying people who are slow to break down acetaldehyde is how unpleasant it makes you feel: flushed, nauseated, rapid heartbeat.[13] So some of the symptoms of a hangover may come from acetaldehyde hanging around.

Consuming high quantities of congeners is thought to make hangovers worse too. There are high concentrations in red wine and distilled spirits, for example, brandy, and low ones in clear spirits such as vodka. Congeners include acetaldehyde, acetones, histamines, and methanol. In fact methanol, a product of sugar fermentation, is thought to be a major contributor to the symptoms of hangover. Alcohol dehydrogenase, ADH, will metabolize methanol at a slower rate than ethanol to form formaldehyde (which is used to preserve bodies) and formic acid (found in the stings of bees and ants), both of which are highly toxic.

2) YOUR NEUROTRANSMITTERS HAVE GONE HAYWIRE I've already explained about the changes in glutamate and GABA, your main neurotransmitter systems. The degree of the imbalance between them—that is, too much glutamate and too little GABA—has in rodents been shown to correlate with the intensity of withdrawal.[14]

3) YOU ARE INFLAMED The inflammatory response happens when your body is damaged, as part of the immune system's response. Despite the fact that it's a natural response, it can be very destructive. Chronic inflammation is now believed to be a significant factor in many long-term health conditions, from diabetes to cancer and, as we shall see in the next chapter, liver cirrhosis.

Alcohol turns on this process because it damages the blood vessels and your gut, so the body then turns on itself. The inflammatory response is unpleasant—symptoms can include nausea, vomiting, headache, confusion, and tremor, as well as clinical depression, which induces mood changes and cognitive impairment, and learning and memory deficits. This is why an anti-inflammatory medicine—for example, ibuprofen—can help the symptoms of hangover.

4) MITOCHONDRIAL DYSFUNCTION Alcohol also damages mitochondrial DNA, particularly in the liver. Mitochondria are the energy-producing machines in every cell and are susceptible to damage from free radicals produced by alcohol via acetaldehyde. Brain cells are reliant on mitochondria for their energy supply, and even slight damage to the mitochondria can lead to toxicity in a number of brain regions.

THE 47 SYMPTOMS OF HANGOVER

In a survey of 1,410 Dutch students, these were the symptoms they named. They're listed in order of how commonly they occurred. So 95.5 percent of those surveyed said they felt fatigued, while only 1.8 percent said they had suicidal thoughts.[15]

Fatigue (being tired)	95.5
Thirst	89.1
Drowsiness	88.3
Sleepiness	87.7
Headache	87.2
Dry mouth	83.0
Nausea	81.4
Weakness	79.9
Reduced alertness	78.5
Concentration problems	77.6
Apathy (lack of interest/concern)	74.0
Increased reaction time	74.0
Reduced appetite	61.9
Clumsiness	51.4
Agitation	49.5
Vertigo	48.0
Memory problems	47.6
Gastrointestinal complaints	46.7

Dizziness	46.0
Stomach pain	44.7
Tremor	38.9
Balance problems	38.6
Restlessness	36.8
Shivering	34.4
Sweating	33.9
Disorientation	33.8
Audio-sensitivity	33.3
Photosensitivity	33.1
Blunted affect (displays less emotion)	29.9
Muscle pain	29.4
Loss of taste	28.0
Regret	27.1
Confusion	25.8
Guilt	25.2
Gastritis	23.4
Impulsivity	22.7
Hot/cold flashes	21.4
Vomiting	20.8
Heart pounding	19.4
Depression	18.9
Palpitations	17.0
Tinnitus	16.8
Nystagmus (involuntary repetitive eye movements)	16.1
Anger	10.1
Respiratory problems	9.7
Anxiety	7.4
Suicidal thoughts	1.8

WHERE DO HANGOVER SYMPTOMS COME FROM?

CAN'T BEAR LIGHT AND SOUND: This is due to the gluta-mate rebound effect.

HEADACHE: May be for the same reason as previously, plus there's also the fact that acetaldehyde dilates the blood vessels in your head. New research also suggests alcohol may cause your immune system to attack your body. It seems alcohol can turn on your immune cells; they are supposed to protect you from bugs and toxins, and they do so by releasing chemicals that cause irritation in blood vessels and nerves, which can lead to pain, including headaches.

There's also an element of dehydration due to the diuretic effect of alcohol (meaning that it makes you pee more). You could try water-loading before going to bed, but of course it disrupts your sleep if you have to get up to pee.

STOMACH SYMPTOMS: Alcohol damages the stomach and intestine lining and so can both give you diarrhea and leave you feeling nauseated. There may be a little inflam-mation in your pancreas too—pancreatitis, inflammation of the pancreas, is commonly associated with heavy drinking.

HOW TO PREVENT A HANGOVER

1. DRINK LESS. You know this already.

2. PACE YOURSELF. It takes about 1.5 hours for you to process one standard drink. Drink slowly and you will have a chance to process the alcohol, which means your blood alcohol peak won't be as high, which has been shown to give less of a hangover.

3. DRINK WATER WITH YOUR ALCOHOLIC DRINKS. Other ways to dilute are to add more ice, add soda to wine or lemonade to beer, or have more mixers with spirits. Nota bene: flat is better as fizzy drinks may make you absorb alcohol faster.

4. DRINK SMALLER MEASURES. Shrinking your glass means less booze over the course of an evening.[16]

5. DRINK CLEAR SPIRITS. They contain fewer congeners, which are thought to make hangovers worse. One study showed that vodka produced a lesser hangover than whiskey, for example (although both the vodka and whiskey drinkers' sleep and next-day concentration were bad).[17]

HANGOVER CURES: WHAT'S THE EVIDENCE?

You can't "cure" a hangover. All you can do at this point is treat the symptoms.

1. Anti-inflammatory. Take ibuprofen when you go to bed and/or when you wake up, ideally with food, as it can be harsh on the stomach lining.
2. Beta blocker. (Nota bene: these need to be prescribed and they are NOT licensed to treat hangovers!) This will slow down the pounding feeling in your head by reducing your heart rate. Whether they do more than that, I'm not sure.
3. Hair of the dog. This is the very worst thing you can do. You are only delaying your hangover by putting alcohol back into your brain. No stars!
4. Hydration. A glass of water before bedtime and one when you get up. Electrolyte/sports drinks also

restore salts; there's no research, but the theory makes sense. * I wouldn't recommend getting hooked up to an IV hydration bag unless you're in hospital, though.

5. Food. You'll likely wake up with low blood sugar, which is why eating carbs makes you feel better. * Eggs contain a lot of the nutrients you need at this point.

6. Caffeine. Any source of caffeine—if you can stomach it—will help you feel more awake. Coca-Cola (flat or not) provides three things that may help: carbs, fluids, caffeine.

7. Probiotics. Probiotics are known to reduce leaky gut, and a very small study on heavy drinkers showed the subjects had reduced levels of inflammation in the body.[18] Promising.

8. Herbs/supplements. Most of these have no evidence. It may be worth taking a B vitamin complex. * Some companies—such as Agrip—are now developing all-in-one combination treatments to speed up alcohol metabolism (see agripdrink.com). Expensive but promising.

9. Move. In theory, any kind of exercise will help you speed up your metabolism and so help shift your hangover. However, you are unlikely to feel like doing this.

WHY YOU SHOULD BE HAPPY IF YOU GET A HANGOVER

Sensible drinking is about restraint, which may become easier when you know that alcohol impairs you the next day. In the 1980s, I used to run the National Institutes of Health inpatient research ward on alcohol just outside Washington, DC.

What was interesting was that, when I interviewed the alcoholics we treated and researched there, almost none of them had ever experienced a hangover. So you could theorize that they'd never had a deterrent to drinking too much.

However, in an online study of Dutch students ages 18 to 30 who'd recently had a hangover, only a small proportion— 13.4 percent—said that if a hangover cure existed, they would increase the amount they drank. Most—71.6 percent—said they wouldn't increase their alcohol consumption.[19] But of course this study asked the respondents what they thought they'd do, not what they'd actually do.

So, all in all, though getting a hangover may mean you're less likely to become an alcoholic, I hope all the other information in this chapter will help you to see how potentially dangerous too much alcohol can be on its journey through the body and brain.

THE HEALTH HARMS
OF ALCOHOL

WHAT'S A SAFE LEVEL?

IS THERE ANY such thing as a safe level of alcohol? This is a question I get asked all the time.

I'd like to throw that question back to you: What do you, as an individual who is responsible for your own health, accept in terms of risk?

After all, if I told you that if you drank a gin and tonic today, you'd die tomorrow, you wouldn't drink it. You wouldn't take a poison pill, would you?

But what if I told you that, if you drink half a bottle of wine a day for 40 years, it will take a year off your life?

The reality of alcohol is this: drinking raises your risk of health harms. But the impact on your health of your chosen level of drinking is not certain. What is known is on a medical and population level: We know there is a strong relationship between alcohol consumption and dying younger. Heavier drinkers die sooner, from a range of disorders. It's proven that alcohol affects more than 200 of the different diseases identified in the *International Statistical Classification of Diseases and Related Health Problems (ICD)*. Alcohol use is one of the top five causes of disease and disability in almost all countries in Europe. In the UK, alcohol is now the leading cause of death in men between the ages of 16 and 54 years, accounting for over 20 percent of the total. More than 75 percent of liver cirrhosis deaths, 7 percent of cancer deaths, and 25 percent of injury deaths in adults under 65 years of age in Europe in 2004 were estimated to be due to alcohol.[1] And in the US, an estimated 88,000 people (approximately 62,000 men and 26,000 women) die from alcohol-related causes annually, making alcohol the third leading preventable cause of death in the United States. The first is tobacco, and the second is poor diet and physical inactivity. In 2014, alcohol-impaired driving fatalities accounted for 9,967 deaths (31 percent of overall driving fatalities).[2] That's a pretty scary list, right?

That said, on an individual level the picture is in no way clear. Some people who drink will die from a disease, for example, cirrhosis, that is directly linked to their consumption of alcohol. Some people will die earlier because alcohol has contributed to a condition, for example, from having a stroke. Some people who reach a century swear that it's their daily drink that explains their advanced years.

When it comes to all drugs, including alcohol—and in fact all risky activities—less is always safer. Looking at the statistics, if you want to maximize your life, the rule would be not to drink a drop.

The same goes for if you want to maximize your health: don't drink at all—because there are no health benefits to it. But if you want the sociability benefits alcohol brings, it's a different story.

In that case, you need to decide what risks you want to accept, balanced out with the pleasure you gain. The risks are determined by how old you are, your sex, your genetics—but, most of all, how much you drink and how often you drink. What you can do is to work out the dose that gives you the best fun but with a reasonable amount of risk—in your opinion. It's not a one-size-fits-all rule. And it really is your call.

By the time you've read this book, the idea is that you'll be able to say: I have made the decision to drink this amount. I have decided that the benefits more than offset my fear of the harms.

I want you to be able to work out how little you need in order to get the effects you do want.

But first, you need to know what all the risks of drinking are and all the potential harms. That's what this chapter is about: giving you the facts on how alcohol affects physical health.

One of the issues is that I don't think the real risks from drinking have been made clear to most of us. We are told by the US government that safe drinking levels are less than one standard drink per day for women and two for men. This limit isn't arbitrary; it was established by a group of experts based on all the available evidence, so it's very solid. But did you know this: if you stick to these levels, your risk of dying due to an alcohol-related condition is around or a little under 1 percent. The guidelines explain that this level of risk is comparable to other regular risks, such as driving. One percent is the level of risk the experts think is acceptable.

But the question for you is, is that the level of risk you're prepared to accept?

In the past, we've been told that drinking—specifically red wine, but also other types of alcohol—is good for cardiovascular health. This is, of course, an extremely appealing message. But do you perhaps use this as an excuse to drink too much? More recently, the increased risk of cancer from drinking has become clear, and it has also emerged that, on balance, no level of drinking is actually beneficial to health.

One study concluded that even a couple of drinks more than four days a week raises your risk of premature death by 20 percent.[3]

Perhaps you drink more than the recommended limit and want to know what the risks are for you. Perhaps, when you have all the information, you might decide to cut down or stop drinking.

The main thing is to know what you are doing—and that is what I aim to help you do. It's up to you: after all, as I said, there is no level of alcohol consumption that is without risk. However, when you drink low amounts, the risks of health harms are low too.

Another factor that makes the risk to health from drinking so difficult to assess is there are a lot of conditions it can lead to, each of them with its own risk, from liver cirrhosis to various cancers and cardiovascular disease.

Contrast this with smoking, which more than 50 years of data has been shown to be the main cause of lung cancer and puts you at high risk of cardiovascular disease. So while smoking leads to a high risk of two conditions, drinking over the safe limits doesn't give such a high risk of any one condition—but it is involved in many more conditions.

WHAT ARE THE RISKS OF DRINKING?

We all know smoking and being overweight are bad for us, and if we smoke or overeat, we know we shouldn't.

But drinking alcohol also has the same kind of long-term or chronic risk, which means it kills you slowly. The eminent statistician David Spiegelhalter uses the concept of the micro-life to explain how fast it does this.

One microlife is 30 minutes off your life expectancy. If you are a 30-year-old man, you will lose one microlife by smoking two cigarettes, or each day of being 11 pounds overweight, or for every four standard drinks. But it gets worse if you drink more than that because of the exponential rise in alcohol harms—see the table overleaf.

ARE YOU SHORTENING YOUR LIFE?

This is an estimate of the impact that drinking at different levels will have on your life expectancy.

It's a calculation based on averages, which means there's no guarantee it will apply exactly to you. But it's useful to know, as at least it gives some proportion to the amount of risk you are taking.

You'll see that the more you drink, the—much—worse it gets because of the exponential nature of the health harms of alcohol. That is, the harm increases much faster than the amount you're drinking.

Here, the impact of alcohol and other risky things is explained in terms of losing microlives.[4] This is a period of half an hour and is used because 1,000,000 half-hours = 57 years, which roughly corresponds to an adult lifetime of exposure. The following table gives an indication of the amount of life lost by drinking each week at different levels over a period of 40 years. I've included some estimates for smoking and being overweight too.

Intake per week	Life shortened in years
10 standard drinks	0.4 years
15 standard drinks	0.8 years
25 standard drinks (half a bottle of wine a day)	2 years
50 standard drinks (1 bottle of wine a day)	7 years
100 standard drinks (2 bottles of wine a day)	21 years
20 cigarettes a day	8 years
Being 5 BMI units a day over a body mass of 22.5 units, i.e., BMI of 27	2.4 years
Being 11 pounds overweight	0.9 years

In addition, the risks from alcohol get bigger the more you drink. In the figure on the next page you can see that for both men and women the increase is not linear but exponential, that is, the risks get proportionally bigger the more you drink. The arrows show that the impact of reducing alcohol consumption from 100 to 50 grams a day actually decreases the risk of death from 16 to 2. In other words, halving consumption decreases the risk by eight times.

Fig.1. Reducing high-end consumption has major impact on harms

Alcohol intake (gm / day)

Source: https://journalsl.sagepub.com/doi/abs/10.1177/0269881113512038.

Another issue is that the link between drinking and disease isn't pointed out to us by government or doctors nearly as clearly as smoking: there aren't pictures of diseased livers or breast cancer tumors on the label of a bottle of Pinot Noir, for example.

And a complicating factor is that while any level of smoking—from just one cigarette—increases your risk of lung cancer and heart disease, the same isn't true of every condition.

For smoking, the graph that shows risk looks like a straight line, rising proportionally the more you do it. But the relationship between alcohol and risk varies according to each condition. For cardiovascular issues, problems start at two or more drinks a day and accelerate when you drink at a higher level. This relationship isn't a straight line but rather looks like a curve, especially when it comes to heavy drinking (see figure).

The good news is, the shape of the curve means that, if you drink a high amount, cutting down will reduce your risk of harm much more than if you are reducing from a low level (see Fig. 1). So, for example, a reduction of four drinks a day when you're drinking a lot—say from 10 drinks a day down to six—will have a much bigger impact on health than dropping from five down to one.

What this means in practice is, if you do drink a lot, it really is worth reducing it.

The fact that cardiovascular health harms don't start at zero has given us a feeling that you can drink "safely." However, in the twenty-first century, researchers have begun to uncover more about the relationship between drinking and, particularly, cancer, where the harms appear to start at a much lower level. For this and possibly (debatably) for brain health, the more you smoke or drink, starting from any level of intake, the more likely you are to get a health condition.

Furthermore, historically most of us have grown up with the idea of a "safe limit" of drinking as laid down by the government. Those past recommendations were based on what was known of the harms then—mainly liver damage and cardiovascular effects. But this language has left us thinking that if 14 drinks a week is "safe," how can two or three or even five a week more be much worse?

We know more now. To reduce the risk of alcohol-related harms, the *2015–2020 Dietary Guidelines for Americans* recommends that if alcohol is consumed, it should be consumed in moderation—up to one drink per day for women and two drinks per day for men—and only by adults of legal drinking age. This is not intended as an average over several days, but rather the amount consumed on any single day. The *Guidelines* also do not recommend that individuals who do not drink alcohol start drinking for any reason.

Interestingly, despite governments worldwide having access to the same science, guidelines vary. The standard drink size or unit—14 grams of alcohol in the US—varies in different countries from 8 to 20 grams. And low-risk drinking guidelines vary hugely too, from 10 to 42 grams per day for women and 10 to 56 grams per day for men; while weekly guidelines vary from 98 to 140 grams for women and 150 to 280 grams for men.[5]

What factors will affect your risk of health harms?
- How often you drink.
- How much you drink.
- Your general health. The poorer your health for other reasons, the more alcohol may affect you.
- Your age. The older you are, the more vulnerable you become to health harms.
- Your sex. Women and men have different health risks.
- Your genetic inheritance.
- The number of years you've been drinking. Most often,

health issues due to alcohol only begin to show up 10 to 20 years after drinking began.

- The age you began drinking. The earlier you began drinking, the more you may have affected your health.
- Whether your family has a history of alcoholism; if it does, you are more likely to become dependent.

WHAT IS A BINGE?

This means getting drunk, as opposed to just having a fun time with alcohol. It implies being intoxicated to a point of being out of control—which in some people means that they lose control of their drinking, then drink massively more than they intended or is good for them.

The CDC defines binge drinking as a pattern of alcohol consumption that brings the blood alcohol concentration (BAC) level to 0.08 percent or more. This pattern of drinking usually corresponds to 5 or more drinks on a single occasion for men and 4 or more drinks on a single occasion for women, generally within about 2 hours.[6]

Not only does bingeing lead to more problems when drunk—for example, doing things that may have a negative impact on your work, your relationships, and even your life, from things like peeing in public to very serious consequences such as molesting other people—but it can have significantly more health impacts than drinking the same amount over a few days. For example, a study from Sussex University in the UK found bingeing damages the brain more than the same amount of alcohol taken in a more spread-out fashion.[7]

A binge can kill you, even if you don't feel particularly drunk. Women are more likely than men to die from a binge. Being on average smaller means that for any given alcohol intake their blood alcohol levels will be higher. That was the sad story of Paula Bishop who, in 2017 over the course of one day on holiday in the Canary Islands, drank a few small beers, two glasses of wine, and four Irish coffees; it's thought around ten drinks. She had a blood alcohol level of 0.4 percent—five times the drunk-driving limit in the US. But, her husband said, she hadn't appeared particularly drunk.[8] (Nota bene: there's another lesson here—measures can be a lot bigger in other countries.)

Bingeing on alcohol can also put you at risk of being harmed or assaulted. That's because when you are this drunk, you become unaware of what is going on around you.

It's also likely that bingeing makes the drinker more vulnerable to becoming addicted because it produces more extreme effects in the brain. That means the brain has to produce more adaptive changes to compensate. I think these changes may one day be proven to be the start of a slippery slope to dependence on alcohol.

Some people, especially younger people, intentionally binge. They drink to get "hammered" or "wasted," to lose control, to forget their everyday life. But there are other people who binge despite not wanting to. They go out for "just a couple" of drinks and then lose control. These

> people can be helped by treatment with nalmefene, a
> medicine that blocks the endorphin rush from alcohol that
> contributes to loss of control and hence more drinking.

THE MAJOR WAYS ALCOHOL
AFFECTS YOUR LENGTH OF LIFE

LIVER

When people think about alcohol and health, this is what they think of first. Because everyone knows drinking too much damages your liver. In fact, it's not the most prevalent cause of harm from alcohol—that accolade goes to cardiovascular conditions. But it's certainly a very important one, and, sadly, one that's becoming increasingly common.

One feature of liver cirrhosis is that the person's abdomen swells up and collects water. A standard piece of medical student education was to examine these patients with one hand on one side of the abdomen and then tap the other side with the free hand. This tap sent a wave through the fluid, which could be felt by the other hand (we call this a thrill). When I was learning medicine, such patients were quite rare overall but relatively common in France, so our name for alcoholic cirrhosis was "the French disease." Since then, the French government has introduced policies that have significantly reduced cirrhosis rates, whereas recent data suggests a similar pattern in the US too.[9]

Alcohol is a poison, and the liver is in charge of breaking down poisons, as well as other drugs and chemicals. That's why alcohol, like other poisons, can destroy your liver. The problem is, most people don't know they have liver damage until it's very advanced—and possibly too late.

1) FATTY LIVER The first stage of liver disease is called hepatic steatosis, which means fatty liver. The liver gets fatty because it makes fat from the calories in the alcohol and stores it in its cells. Your liver may also be fatty if you are overweight, have insulin resistance, or are diabetic—this type of nonalcoholic fatty liver disease is very common.

One issue is that most people don't feel any symptoms during this early stage of liver damage. You might feel a little abdominal discomfort over your liver—which is in the upper right of your abdomen—perhaps a little heartburn or nausea.

If a GP (general practitioner, also known as a primary care physician) thinks you may have fatty liver, they will do blood tests. You may also be offered a Fibroscan, a specialist ultrasound that can measure scarring and fatty change in your liver.

It's been shown that a lot of middle-aged "average" drinkers have fatty liver. According to the charity Drinkaware, you are at risk if you drink more than six drinks a day if you are a man and five if you are a woman—for just two or three weeks, which is only as long as a long vacation! NHS (National Health Service, the UK's publicly funded healthcare system) advice says if you stop drinking for two weeks, you can reverse fatty liver disease (after this, you need to carry on drinking within the guidelines, of course).[10] But the time it takes to recover will depend on the severity of your condition, among other factors, especially being overweight.

2) ALCOHOLIC HEPATITIS Most of the alcohol you drink is broken down in the liver, into energy plus acetaldehyde, which at one carbon atom longer than formaldehyde (used to preserve Damien Hirst's art exhibits of sliced sharks and cows) is even more of a toxic pickling agent than alcohol. This poison also kick-starts the process of inflammation.

If you go on a significant binge—maybe half to a whole bottle of spirits or seven drinks in the course of an evening—you can get acute alcoholic hepatitis. Your liver will become inflamed by the vast amount of alcohol you have taken. Occasionally people die of a very fatty liver that becomes inflamed.

It's more common to develop acute alcoholic hepatitis over three to four days of drinking. We now see this condition in young binge drinkers; in fact, there are wards full of these kids. There is a particular danger point for dying the night you become a legal drinker—18 in the UK, 21 in the US—because people press drinks on the young person, thinking they are being nice. For the parents of these children, it becomes the worst day of all.

Binge drinking is particularly dangerous because when the liver dies, you tend to die. The only way to recover from it is by having a liver transplant. However, bingeing is not the reason behind most cases of alcoholic hepatitis; it is diagnosed much more often in long-term heavy drinkers.

There's also another type of hepatitis that occurs in drinkers: autoimmune hepatitis. When your liver starts to get damaged by alcohol, and inflamed, it becomes more vulnerable to attack by the immune system.

Finally, if you have some other form of hepatitis that's not caused by alcohol—for example, hepatitis A, B, or C—then drinking alcohol massively increases the rate of dying from this.

3) CIRRHOSIS This is the most serious form of liver damage. The liver has quite a lot of regenerative capacity—which is why you can recover from fatty liver—but there comes a point of no return. And cirrhosis is it.

In the same way you might pickle fresh fruit in alcohol to make the German pudding *rumtopf*, alcohol pickles your liver. Cirrhosis happens when, over decades, the mixture of alcohol and acetaldehyde permeates the tissues of the liver and begins to freeze all the metabolic activity. The tissues can't even degenerate because the enzymes that usually digest the dead cells, the ones that are also responsible for the breakdown of a dead body, are killed too. In this state, the liver gradually solidifies.

As you become more tolerant to alcohol and the amount you drink increases, more and more of the liver dies, so that instead of being a soft, squidgy organ that's full of blood, it becomes dry and hard and woody.

It loses its capacity to make the things you need it to make, such as enzymes and some hormones. In fact, when your liver starts to pack up, your alcohol tolerance disappears because it can no longer metabolize the alcohol. That's when you start to get drunk on small amounts of alcohol. If it becomes clear this is what is happening, you must stop drinking immediately. You are facing terminal liver failure.

As the liver becomes more and more blocked by this fibrous tissue, it becomes harder for blood to flow through it. As most of the blood from your gut goes through the liver, it has to go somewhere, so it starts going through little vessels in the walls of the gut and the gullet (esophagus), which start to carry far more blood than they were designed for. These veins get bigger and bulgier. The bulges are called varices and are similar to the bulges found in leg veins when you have varicose veins.

Often, the first time someone finds out they have serious liver disease is when they start vomiting blood when one of these vessels has burst. In the hospital, an operation or a pressure tube in the esophagus can stop the vessels bleeding, but the more serious issue is that by that stage, the liver is usually beyond repair.

Toward the end of the cirrhosis process, you can no longer make the enzymes you need to digest your food. Another key function of the liver is to make the proteins that your body needs to make blood. When your liver isn't working, your protein levels go down and you get the fluid build-up in your abdomen that I described above. It's this metabolic disturbance that eventually stops your heart.

If you do have some liver function left, you can stay alive with cirrhosis for some years. But you are extremely vulnerable to catching a fatal infection; your body easily becomes overwhelmed, as you don't have the right level of proteins in your blood.

The other thing the liver does is break down the toxins produced by the microbiome so that they don't get to the brain. That's why people at this stage get confused and delirious—a phenomenon called hepatic encephalopathy—because their brain is being poisoned by these substances. Doctors can give people antibiotics to kill the gut bugs, but this only gives temporary respite. At this point, life expectancy is very short—unless you have a transplant.

Mickey Mantle, a long-term alcoholic, was a classic case of cirrhosis. Not only did he kill his own liver, but he was lucky enough to be given a transplant—leading to criticism that celebrities had a better chance at getting one.

You might be thinking, is it appropriate to give such a scarce and precious resource to a person who may drink again, and so squander it? I think about it like this: the vast majority of people who drink don't end up dependent. And as alcohol is legal, you can't blame the people who try it, then do get addicted. They didn't intend to become alcoholic; they are the unlucky ones. Should you penalize someone for being unlucky? This debate has raged for years, and while in the past doctors required alcoholics to be sober for six months before receiving a transplant, this mandate is now being waived more often due to new studies, leading to a rise in transplants for people with alcohol-associated liver damage (ALD). Recent research shows that survival rates post-liver transplant is only 11 percent lower for people with ALD.[11]

You don't have to be the cliché of an alcoholic who is drinking day and night to die of cirrhosis. A small study by the UK's leading alcohol and liver disease expert, Professor Nick Sheron, found that a third of patients with severe alcohol-induced liver damage had never even considered

that they were drinking abnormally. They met the criteria
for excessive drinking without ever having realized it. Only
9 percent showed evidence of severe alcohol dependence.
Most had not lost control of their drinking; rather they
were classified as heavy drinkers.[12] Professor Sheron says:
"The majority of patients presenting with alcoholic liver
disease appear to be heavy controlled or social drinkers,
leading relatively controlled lives, perhaps not feeling that
their drinking is a major health issue until they are diagnosed with end-stage liver disease, at which point the liver
has been damaged to the extent that only 30 per cent will
be long-term survivors."

We really do not understand why some people get cirrhosis and others do not. We presume it's something to do with
the body's immune response and genetics and is likely to be
aggravated by poor diet.

Patterns of drinking seem to matter too. According to a
UK-based initiative to explore the factors that predispose to
breast cancer, daily drinkers have a higher risk of cirrhosis
than non-daily drinkers, and women who drink mainly
with meals have a lower risk than those who drink without
food.[13] It also seems that women may be more susceptible
to the toxic effects of alcohol on the liver.[14]

But the important message for everyone is: you don't
have to be dependent in the sense that you crave alcohol or
have trouble stopping drinking in order for it to kill you.

4) LIVER CANCER Any chronic inflammation can lead to
cancer. So one of the other possible consequences of cirrhosis is primary liver cancer—cancer that starts in the liver. If
you have cirrhosis, you have a 12 percent lifetime risk of

developing liver cancer. It is very hard to treat and is usually a terminal diagnosis.[15]

OTHER CANCERS

This is the—pretty long and scary—list of other cancers for which alcohol is known to be a significant risk factor: breast, colorectal, esophageal, pharynx and larynx, lip and oral cavity, and nasal. Some cancers are dose dependent (see box on page 56), but some aren't.

The causal association is strong enough that alcoholic drinks have been classified as a carcinogen by the International Agency for Research on Cancer. And it's also why the World Health Organization has decided that no level of alcohol consumption is safe.

That doesn't mean there aren't other cancers where alcohol contributes. For example, World Cancer Research Fund says there is limited but "suggestive" evidence that drinking may contribute to cancers of the lungs, pancreas, and skin.[16] And we're finding out more all the time; for example, it was only in 2017 that the full extent of the association of drinking and breast cancer became clear. According to a World Cancer Research Fund report from that year, just a single drink a day increases the risk of breast cancer.[17]

There is some good news: a moderate intake of alcohol (under two drinks a day) appears to lower the risk of kidney cancer .

There are two ways in which alcohol increases the risk of cancer. First, there is the acute toxic effect of alcohol on the skin of your mouth and gullet, where it burns and damages the tissue. This chronic inflammation leads to cancer through changing DNA, particularly once it has led to persistent mouth ulcers that last for months. While it's not known why some people go on to develop, for example, mouth cancer, it is known that drinking is a major contributory factor (others are smoking and the human papilloma virus or HPV).

The second way is by a more indirect route, which will vary for each type of cancer. This process is thought to involve the production in cells of molecules called free radicals, which then leads to the DNA damage that promotes cancer.

DRINK AND CANCER: HOW MUCH ALCOHOL CAUSES CANCER?

There is strong evidence that consuming

- Any alcohol at all increases the risk of mouth, pharynx, and larynx cancers, esophageal cancer (squamous cell carcinoma), and breast cancer (pre- and post-menopause)
- 2 or more drinks a day (30 grams or more) increases the risk of colorectal cancer
- 3 drinks a day (45 grams or more) increases the risk of stomach cancer and liver cancer
- 2 drinks a day (up to 30 grams) decreases the risk of kidney cancer[18]

How much alcohol do you need to consume to increase the risk of cancer? This statement comes from AMPHORA, a recent, very detailed European Commission report on the many effects of alcohol.[19] Nota bene: This statement also refers to all the other carcinogens in alcoholic drinks. But as they're present at levels much lower than ethanol itself, it's the ethanol and the acetaldehyde[20] that are the main carcinogens.

Ethanol is a carcinogen, a teratogen, and toxic to many body organs. Using the European Food Standards Authority guidance on risky exposure for human consumption of toxic substances in food and drink products, drinkers consume more than 600 times the exposure level for genotoxic carcinogens, which is set at 50 milligrams alcohol per day (.05 grams/day), and more than 100 times the exposure level for noncarcinogenic toxins, which is set at 0.3 grams alcohol per day.

This means that to stay below the acceptable food standard risk for cancer you should consume no more than 50 milligrams per day. This works out as 18 grams per year, that is, a maximum of two standard drinks a year!

CARDIOVASCULAR HEALTH AND BLOOD PRESSURE

There are a few ways in which alcohol affects cardiovascular health. Drinking can damage the heart, a condition called cardiomyopathy. It weakens the heart muscle so it becomes less efficient, less powerful at pumping the blood. It is really poisoning your heart in the same way as it poisons your liver. Over many years, this can eventually lead to heart failure.

Even light to moderate drinking raises your risk of an

irregular heartbeat, cardiac arrythmia, which may make you feel faint and short of breath.[21]

More important, alcohol is also the leading preventable cause of hypertension (high blood pressure). According to US figures, drinking contributes to at least 7 percent of cases. Why is increased blood pressure dangerous? Because it leads to two main outcomes: heart attacks and strokes.

We don't fully understand why drinking raises blood pressure. One reason might be that during alcohol withdrawal— which, you will remember, you go through every time you drink—there's a surge in the neurotransmitter noradrenaline, and that's known to increase blood pressure.

It used to be thought that alcohol provided some protection against stroke. In the past, this has been the source of countless headlines about the benefits of red wine. Because everyone who drinks loves to read all the reasons why it's a good idea, right?

There's now controversy over whether alcohol does provide any protection. A 2018 review in *The Lancet*, one of the leading medical journals, was called—pretty definitively—"No Level of Alcohol Consumption Improves Health."[22] Their conclusion was that, on balance, any protection would be more than canceled out by the negative effects.

The Lancet article has been criticized for stating this as fact, when the way the studies are set up—they are observational—means they can only say there is a correlation between a level of drinking and a disease, rather than a cause. It also didn't differentiate between what people were drinking: whether it was wine, lager, cider, or vodka, for example.[23]

The critics of this report aren't saying that heavy drinking isn't bad for you: that is well established. But what we don't know is what happens to the health risks between no drinks and one or two daily drinks—and one reason for this is that the difference in risk is so small.

There may well be a partial protective effect on cardiovascular health—*The Lancet* paper suggests this is true for women, at least, though previous research has pointed out this effect in men too. But there are two really important points to mention. The first is that the amount of alcohol optimal to provide the protection appears to be very low—about half a standard drink a day. Second, that benefit to the heart does not outweigh all the other risks of alcohol, most notably cancer.

When it comes to alcohol, it's been shown that the risk of stroke is both dose dependent and cumulative.[24] I don't think most people who are being treated for hypertension are even asked whether they're drinking too much; I suspect because doctors feel uncomfortable asking this—perhaps because many doctors drink too much themselves?

There is some good news. It seems the effect on blood pressure is reversible if you stop drinking. In one study,[25] 13 out of 18 heavy drinkers saw their blood pressure return to normal levels after one month of abstinence. However, if you're a heavy drinker, you'll need to cut down slowly rather than cut it out immediately, or you may get a rebound effect of very high blood pressure from the noradrenaline surge that comes with withdrawal. This rebound increase in blood pressure during withdrawal explains the finding that bingeing is significantly more likely to result in a stroke.[26]

It also appears that moderate drinking may not be a risk factor, for blood pressure at least. A review in *The Lancet* in 2017 concluded that if you drink two or fewer drinks (24 grams of alcohol) per day, you'll see no significant effect on blood pressure from drinking less.[27] However, anyone who drinks above that amount will see benefits if they cut down. And the more you drink, the greater the benefits of cutting down.

BRAIN

This may shock you, but drinking causes brain damage. In fact, the leading preventable causes of dementia are head injury and the damage alcohol does to the brain.

It's thought at least one in five cases of dementia is probably due to alcohol. And so a lot of people will be affected: As you probably know, dementia is one area of medicine where there are a growing number suffering. In fact, the number of people living with it is expected to triple by 2050. And it's been suggested that women may experience brain damage at lower levels of alcohol intake than men.

In a 30-year study—the Whitehall II Imaging Study[28]—which involved over 500 people having their brains scanned and their drinking histories taken, there were increased odds of hippocampal atrophy in those drinking up to 10 drinks weekly, and faster cognitive decline in those drinking up to five drinks weekly. The conclusion? Alcohol consumption, even at moderate levels, is associated with adverse brain outcomes.

As always, there is contradictory evidence, from a cohort study that looked at the drinking habits of 9,000 mainly middle-aged civil servants, and followed them for 23 years. It showed that low levels of alcohol consumption—that is between one and 10 drinks a week—reduced the risk of dementia. In fact, it appears that being teetotal may raise your risk of dementia—though drinking more than 10 drinks does too. However, the authors do say that you shouldn't start drinking if you don't already, due to all the other health harms.[29]

What is absolutely certain is that heavy drinking affects the brain. A couple of years ago, there was a man in our local neurology outpatient clinic being assessed for dementia because he was forgetful, agitated, and confused. He had been tried on various, pretty expensive, medications for dementia but nothing seemed to work, and they were struggling to know what to do with him. Then, one of my trainee psychiatrists took over his case. He realized nobody had ever taken a drinking history. It turned out this man was drinking huge amounts every night, so he was in alcohol withdrawal every day. When we helped him stop drinking, his symptoms got significantly better.

There are several reasons for the impact of alcohol on the brain. There's the direct neurotoxic effect of the alcohol itself. There are indirect effects, such as the increased risk of stroke, diabetes, and high blood pressure that comes from drinking. All of these raise your risk of dementia. Finally, you're more likely to have an accident after drinking—and head injury is a huge risk factor for dementia.

There's a famous condition that revolutionized our understanding of what happens to the brain in drinkers: Korsakoff's syndrome. The main symptom of this is losing the ability to lay down new memories. I've had patients come into my clinic who I've spoken to at length, and then the next time they come in, they have no recollection of ever seeing me before—even though they can recall distant memories, from their childhood or where they went to sea or fought in one of the world wars. People with this syndrome can often be misdiagnosed as playacting. It's usually found in very heavy drinkers, and it's caused by a combination of excess alcohol and vitamin B1 (thiamine) deficiency damaging a key part of the brain that's vital for laying down new memories.

There is also an intermediate stage on the way to this, called Wernicke's encephalopathy. This is an acute inflammation of the brain characterized by a state of complete confusion—delirium, double vision, being very unsteady when standing. At this stage, if you treat the patient with thiamine (vitamin B1), you can prevent them from ending up with brain damage.

That's why, for almost a century, some doctors have been saying that all alcoholic drinks should have added vitamin B1. I'd support this. Although people with acutely inflamed brains are relatively rare, it's likely there are a lot of people walking around with milder forms. That is why when anyone seeks help with their drinking, we give them high doses of vitamin B1—up to 1 gram per day.

WHAT IS THE FRENCH PARADOX?

The term French Paradox was coined in 1980. It described the fact that French people had lower levels of heart disease than other nations while consuming a high level of saturated fat—think Camembert and cassoulet—and as much alcohol. (Nota bene: the French didn't have a lower level of liver damage or cirrhosis.)

The theory was, the missing link was red wine. In other words, that consuming this allowed the French to eat cheese and meat while keeping their arteries clear. The idea was that red wine was cardio-protective, but that spirits and beer were not. One chemical that occurs naturally in wine, resveratrol, came under particular scrutiny. But its early promise of being a magic cardio-protective bullet didn't come to fruition.

There have been big claims made about what resveratrol can do, including preventing both cardiovascular disease and cancer. That's why you'll see it on sale in health food stores. But most of the claims are based on *in vitro* and animal studies. And the amount you'd need to take to see any of these benefits, for the most part, is far larger than the amount you'd get in a glass of red wine.[30]

The real reason for the French Paradox is likely not wine, but the mainly Mediterranean diet the French were consuming, which was rich in olive oil and contained more fish and chicken than red meat, as well as lots of fruits and vegetables, wholegrains, and legumes. And this is now proven to be good for heart health. There are other factors that weren't taken into consideration besides wine: more sun and so more vitamin D, as well as the more relaxed way that southern Europeans tend to eat. There's also the fact that the French, along with other southern Europeans, drink differently than northern Europeans. They're less likely to binge, more likely to drink consistently. (Nota bene: this kind of drinking with no days off may be worse for the liver.)

Fig. 2. The more you drink, the more diseases you get: cause specific relative risk by alcohol consumption

— Men and women
-- Men
···· Women

Source: https://www.bmj.com/content/325/7357/191/related.

This figure shows the impact of increasing levels of drinking on the risk of a range of different medical disorders on both men and women.

Evidence does suggest that drinking small amounts of red wine might possibly be a little cardio-protective. If you look at the figure you will see that in the top right-hand box, there is a slight downward trend for ischemic heart disease in men. You'll also see it is the only downward trend out of any of the conditions. But the optimal amount of alcohol to consume per week to get this protective effect is about equivalent to just 75 milliliters of 13.5 percent ABV wine (less than half a 5-ounce glass in a bar). Drink any more, and the effects become negative. When consumed in that volume (or lack of it), you're really treating red wine more like having a shot of medicine rather than a recreational drink.

No one has done an experiment that has proved definitively

that drinking alcohol—red wine or any other form—is protective for cardiovascular health. All the claims are by association, not causation, and apply to the specific population studied. So we can't say for certain that any given individual will benefit. We will probably never get definitive proof—it's very expensive to do the kind of study that could establish this.

There has been one recent attempt to set up such a study. The Moderate Alcohol and Cardiovascular Health Trial, a six-year randomized controlled trial of nearly 8,000 over-50s in 16 global sites, had its extensive funding raised by the US National Institutes of Health's National Institute on Alcohol Abuse and Alcoholism. The trial proposed that one group of subjects would be abstinent, the other would have one drink a day. This would finally answer the question: What effect does a low level of alcohol have on cardiovascular health—the rate of heart attacks, strokes, and death?

The problem was, the cost—over $100 million. The alcohol industry agreed to provide funding, which in theory wouldn't influence the actual study itself. But then all hell broke loose. There were those who objected because the study wouldn't include analysis of some possible negative effects, including breast cancer. But the main objection was that funding by the alcohol industry would skew the findings toward alcohol having a positive effect. Some believed alcohol companies had had too much contact with the scientists involved. And there were those who thought that the researchers shouldn't take money from companies that made it by harming people. In the end, after an investigation by the US senators who regulate NIH funding, the study was dropped.[31]

DIABETES

Like cardiovascular health, there has been some evidence to suggest that drinking lightly or moderately may protect against type 2 diabetes. But also in a similar way, there is debate: a large-scale review from 2015 suggested that any beneficial effects may have been overstated and that if they do exist, they may affect only women and non-Asian populations.[32] However, a 2017 review concluded that light to moderate drinking does reduce the risk of diabetes.[33]

What doctors do agree on is that heavy drinking increases the risk of type 2 diabetes.

Alcohol inflames the pancreas, the main function of which is to secrete digestive enzymes into the small intestine. These are surprisingly powerful, and what seems to happen is that once the pancreas is damaged by alcohol, it starts to digest itself, which is very painful.

This effect is dose related, and usually happens after 10 to 20 years of drinking. Women seem to be more vulnerable than men. We know that alcohol is the most common reason why people get pancreatitis. And once the pancreas is damaged, this can lead to insulin-dependent diabetes.

A second factor that increases the likelihood of diabetes is the calorific content of alcohol. Drinking a lot increases your total calorific load, and being overweight can push you in the direction of type 2 diabetes.

If you already have diabetes—either type 1 or 2—alcohol can affect you in other ways. Alcohol disrupts your sugar balance and metabolism, leaving you more vulnerable to both hypoglycemia and hyperglycemia (low and high blood sugar levels). And as alcohol affects your brain function, you might not notice what's going on or be less likely to maintain your proper insulin regime or take steps to sort out your blood sugar.

Another big concern is that a hypo, a state that occurs when blood sugar levels drop below those needed to keep your brain working properly, can look like being drunk, so you may not get the right treatment. The thing to do here is to wear a medical ID bracelet, so people know you have diabetes.

THE GUT

One very common—though not life-threatening—side effect of drinking is gastric reflux. This is when the sphincter between your esophagus and stomach opens when you lie down at night, which isn't supposed to happen. The result is that stomach acid comes up into your esophagus and burns it, which can be painful. It's especially common if you are partial to beer, due to the large fluid intake. But it's also due to the fact that alcohol delays gastric emptying and can stimulate gastric acid secretion. Drinking neat spirits will also burn and inflame the esophagus.

The stomach itself seems to be relatively resistant to alcohol, presumably due to the coating that stops it from being destroyed by stomach acid. There is an exception: a stomach ulcer. These are caused by bacteria called helicobacter, and alcohol makes it easier for the bacteria to bed in. Ulcers are not generally life-threatening, unless they perforate, but they can turn malignant and so lead to stomach cancer.

Alcohol is also known to impair the absorption of nutrients in the small intestine. This is especially true of vitamin B1, the malabsorption of which leads to brain damage (as discussed earlier). It may also affect uptake or absorption of folic acid, thiamine, B12, zinc, and selenium.

Some alcoholics, those who can no longer look after themselves or who buy alcohol instead of food or who live on the street, are vitamin deficient in any case, due to their very poor diet. I have in the past described their diet as being just French fries, which may sound flippant but it can often be true. If a patient these days presents with scurvy, which is caused by vitamin C deficiency, they are usually an alcoholic.

Drinking makes the protective membrane of the gut wall more permeable too—a condition called leaky gut. Leaky gut also allows toxic substances produced by the gut bacteria into the blood, and this can lead to brain toxicity. Normally the liver mops these up, but in liver failure it can't do this adequately and a state called hepatic encephalopathy ensues. One of my patients in the late 1970s was a woman dying of end stage cirrhosis—it wasn't due to alcohol but from hepatitis B. Over the years it had destroyed her liver, as alcohol can too.

She'd become confused and delirious, and the reason was that the toxic substances from her gut bacteria weren't being cleared by the liver so they got into her brain and poisoned it.

We showed this was the case when we treated her with antibiotics. They kill all the bacteria in the gut, all the microbiome, so the toxins disappear. For a few days, she woke up, was alert and conscious. But then her gut recolonized itself and she went back into a coma and died.

Right now, the microbiome is a hot area of research. The suggestion is that your unique inner world of bugs is not only affected by what you drink (and eat) but is a key factor in whether you get ill and with which condition. Drinking alcohol (and/or having a bad diet) encourages the microbes in the gut to produce toxic substances that promote the inflammatory processes and gut permeability that allow diseases, including liver disease and cancer, to happen. This means the gut may also turn out to be where we can look to treat or prevent these conditions.[34]

AAA: ALCOHOL, ACCIDENTS, AND AGGRESSION

EMERGENCY DEPARTMENTS ON weekend nights are testimony to the damaging effects of alcohol. The medical staff dread the night shifts, especially today when it seems to me to be more acceptable to abuse health-care staff than it was when I was a junior doctor. In my memory, most of my drunk patients were good-natured, even apologetic. Now they seem to be much more frequently aggressive and violent.

On top of chaos in the emergency room, alcohol-related community hospital admissions in the US have increased over the past two decades, and it has been estimated that these now total around 400,000 each year.[1]

ACCIDENTS DO HAPPEN

And they happen more frequently when you're drunk. Have you ever woken up from a big night out with a bruise and you don't know how you got it (aka an unidentified drinking accident)?

That's because at low doses, as well as making you clumsy, alcohol luckily has a pain-reducing effect, which seems to be mediated through turning on parts of the GABA system.

Then as you push the dose up and it affects the glutamate system, that's when the anesthesia effect kicks in (it's also when the amnesia kicks in too, which is why you can't remember how you got that bruise/break/bump). It's a not-uncommon sight in emergency departments to see people who've broken bones and are in massive pain but are still standing. Although they may not be able to remember, or talk sense.

IN FACT, PEOPLE HAVE ACCIDENTS AND DIE

People do a lot of stupid things when they are drunk. It's a lethal combination: Alcohol helps you shed your fear but also your judgment. You think: *Oh, I can reach that*—when obviously you can't. When drunk, people are also more likely to succumb to dares and challenges. I was going to write "stupid dares," but the truth is, all dares when you are drunk are stupid.

Think of all the vacationers we hear about, every year, who fall and die in Benidorm and Magaluf, trying to jump into swimming pools from heights, or climb from balcony to balcony. Death and injures are so common in these Spanish resorts—especially among Brits—that this ridiculous activity has been given its own verb: "balconing."

No matter where they are, people die from climbing buildings and falling off them. Or they die from jumping into the sea from piers and drowning—if the water is cold, your heart can stop as you hit it.

DRUNK DRIVING

When I describe a car in the hands of a drunk person, I liken it to a massive weapon. In 2016, 10,497 people died in alcohol-impaired driving crashes, accounting for 28 percent of all traffic-related deaths in the United States. Looking beyond the statistic, every one of these accidents is a personal tragedy for someone's family and friends. Sometimes four or five young people all die in one crash because the driver is drunk.

That's not including the many thousands of accidents where people don't die. More than 2.5 million Americans went to the emergency department (ED)—and nearly 200,000 were then hospitalized—for crash injuries in 2012. On average, each crash-related ED visit costs about $3,300 and each hospitalization costs about $57,000 over a person's lifetime.[2]

You won't be surprised to hear that young adults (ages 16 to 24) are overrepresented in drunk-driving casualties, making up a quarter of them (compared to one in five of all accidents). You may also be unsurprised that four out of five drunk-driving accidents involve male drivers. Looking at the list of famous people who've been arrested for drunk driving, it is overwhelmingly male: Mel Gibson, Justin Bieber, Michael Phelps, and Tiger Woods. This isn't to shame them. Being dependent on alcohol is a brain disease, not a failure of will.

Compounding the problem is that men tend to overestimate their capacity to drive at the best of times, and this overconfidence is made much worse when they are drunk.[3]

And let's not forget that a lot of pedestrian accidents are caused by being drunk too—because people walk in front of cars. Out in Oxford one night when I was starting my research career, I saw a man staggering down the pavement. Suddenly he walked into the road—and got hit by a car. He bounced off the car and fell over. Then he got up and staggered away, ignoring the driver's offer to check if he was OK. He was so drunk, he didn't feel the pain.

There are three reasons drinking is really bad for your driving.

1. It affects your judgment and makes you think you are a better driver than you are. Because alcohol takes away the higher-level control of your conscious brain, you lose insight into what is really going on, as well as your ability to make a rational decision.

2. Your coordination is much, much worse. Under the influence, some people drive faster and some slower. The latter can happen when drivers know they are drunk and try to overcompensate. If you have ever followed someone home from a bar after closing time and they've been driving at 20 mph, you can probably assume they have had too much to drink.

   Your reaction times are also slower. So if something happens—a cat runs out onto the road, or another drunk person—you're not as quick at responding.

3. You're more likely to fall asleep at the wheel. Alcohol can

make you much more sleepy, especially when you're tired or you're driving at night.

DO YOU KNOW YOUR LIMITS?

A new national survey shows most Americans (63 percent) do not know the legal limit blood alcohol concentration (BAC) for drunk driving, which is the same across all 50 states and the District of Columbia.[4]

In the US, the limit is 0.08 percent unless you are under 21, when it ranges from zero to 0.02 percent (effectively zero). The limit for commercial drivers is lower because they drive more and the harms they would cause from having an accident are so much greater. Some states in the US still have functional testing, similar to how we tested drivers in the UK pre-Breathalyzer. This involves making the suspect prove that they are not intoxicated by having them perform a number of psychomotor coordination and memory tasks. These include toe-to-heel walking in a straight line, following a moving finger, and recalling spoken sentences, all of which are impaired by alcohol. If you fail the test, you can still be prosecuted even if you are under the breathalyzer limit.

You could argue functional testing is fairer, as it shows if you are capable of driving—or not. That said, even if you are physically coordinated, your risk-taking faculties might be seriously impaired. The other issue with functional testing is that it doesn't pick up whether you might fall asleep at the wheel.

In 1981, the introduction of an accurate Breathalyzer meant police could test drivers by the side of the road. The maximum breath alcohol concentration is 35 micrograms of alcohol in 100 milliliters of breath (this is the equivalent of 0.08 percent in your blood).

If you've ever been Breathalyzed, you'll know what the form is: The police give you a preliminary breath test at the side of the road. If you are over the limit, you can be taken to a police station for an evidential breath, blood, or urine test. If you are convicted of drunk driving, you will almost certainly be banned from driving and be given an unlimited fine. In extreme cases you may even be sent to prison.

HOW MANY DRINKS CAN I HAVE AND DRIVE?

Do you know what the 0.08 percent BAC limit means for you? According to the government, it's impossible to say exactly how many drinks will take you to the limit, as it's different for each person. It will depend on your weight, age, sex, and metabolism, as well as what you were drinking and what you have eaten.

The problem is, people tend to assume they know how much takes them over the limit. Some—very sensibly—decide not to drink at all when they drive.

The rule of thumb is that an average-sized man can drink up to two drinks of alcohol, and a woman up to one and a half before they exceed the US legal limit. However, who is average? Unless you have actually been Breathalyzed, it's extremely hard to know how fast your body eliminates alcohol—and you'll never know exactly how fast you're processing it on any one particular day.

I suspect most people over-guesstimate how much they can drink. You could work out how much you can drink according to the fact that you process about two thirds of a standard drink per hour; but this, I think, makes for confusion. Not only do you first have to count the amount you have drunk, but then you also have to count your drinks from the time you started, and then count the time after you stop drinking. The possible margin of error is wide, not least because alcohol affects your ability to pay attention to these details. Along the same lines, people who estimate their blood alcohol as lower than it is tend to drive more riskily.[5]

There are various unit calculators online, where you input your sex, weight, age, and so on, but they all state very clearly that they aren't accurate and that they should only be used the morning after.

I was told by one man that he always drank three pints and drove because he had, and I quote, a "fast metabolism." And by a woman that two glasses of wine is fine if it's with dinner. But plugging this into a blood alcohol calculator, for example, if she drinks two 5-ounce glasses of 13 percent wine, it will take six hours to leave her system completely.[6]

Fig. 3. The effect of drinking on risk of a road traffic accident

Blood alcohol concentration (mg/100 ml)

Source: www.ncbi.nlm.nih.gov/pubmed/18955613.

AAA: ALCOHOL, ACCIDENTS, AND AGGRESSION

The point is that the 0.08 percent BAC is a legal threshold—it's not a safe level. The figure shows that the risks of an accident start to rise almost from when you start drinking, that any level of alcohol can impair driving to some extent. At the 0.08 percent limit, on average, your likelihood of having a road traffic accident is at least three times more than if you had stayed sober. For those most susceptible, for example, young people, the risk may be up to five times greater.

And having alcohol in your blood has an even greater impact on whether you die as a result of a road crash.[7]

"How many drinks can I have and drive?" is not the question you should be asking yourself. Because it's not about staying within the legal limit, it's about staying safe.

I'd advise you, if you are driving, not to drink at all. This table, which shows how blood alcohol concentration affects the likelihood of you having an accident, might convince you:

BLOOD LEVEL/RISK OF ACCIDENT

These figures are for young people under 19—for older people the risk rates do come down quite significantly. But 0.05 percent is significantly lower risk than the current 0.08 percent for all age ranges. Also note the extremely steep rise in risk once blood levels get over 0.1 percent—this applies to people of all ages.

- 0.02 percent was 2 times the risk
- 0.05 percent was 3 times the risk
- 0.08 percent was 8 times the risk
- 0.1 percent was more than 12 times the risk[8]

What we are not told—when we carefully measure out our drink or drinks that we think take us up to the drunk-driving limit—is that driving with a blood alcohol concentration of 0.08 percent is risky.

Even in 1967—more than 50 years ago—when the level was first set, experts knew that people driving at this level were under "significant impairment." The figure 0.08 percent was only chosen as an "acceptable compromise" at the time, when it was felt the public was against any limit at all, despite the high number of road traffic accidents at the time.

The current limit gives the impression that it is safe to drink a glass or two, but that is not the case. If being "significantly impaired" is not a reason not to touch a drop before you drive, I don't know what is.

The *Report of the Review of Drink and Drug Driving Law* by Sir Peter North was commissioned by the UK government and published in 2010. It is a comprehensive review of drunk driving, and it states that if you are driving and your blood alcohol concentration is between 0.05 and 0.08 percent, you are six times more likely to die in a collision than if it was zero. And you are 2–2.5 times more likely to be involved in an accident than drivers who've drunk no alcohol.[9]

The most important message is, being under the threshold doesn't mean you're fine to drive. The difference between 0.079 and 0.081 percent is not that one is safe and the other dangerous, but that one is legal and the other is not.

New neuroscience research from Sussex University in the UK backs this up. It shows that even one pint of beer can compromise driving safety. Even when your behavior and coordination are not affected, after just two drinks your feeling of being in control of your actions is exaggerated. And the resulting overconfidence in your driving ability could be dangerous.[10]

In many ways, a zero limit is the most rational approach (although I suspect the legislation would never get through); at zero, there is no temptation to guess or play the system.

SHOULD THE DRUNK-DRIVING LIMIT GO DOWN?

In 2014, in response to the fact that Scotland had so many alcohol-related road accidents, the Scottish Parliament lowered the drunk-driving limit from 0.08 percent to 0.05 percent for blood (which is 35 micrograms down to 22 micrograms per 100 milliliters of breath).

This puts Scotland in-line with almost every other country in Europe, except Malta. Germany, Italy, Austria, Croatia, Belgium, Greece, Ireland, Portugal, Spain, and France are all 0.05 percent. Estonia, Poland, Sweden, and Norway are all 0.02 percent—which in practical terms means you cannot drink at all. The tiny amount takes account of the fact that some foods contain small amounts of alcohol, and that the body and the gut bacteria produce alcohol from food. The Czech Republic, Hungary, Romania, and Slovakia have a limit of zero[11] (even though in reality, as explained above, it can never actually be zero).

I'm all in favor of taking the limit down from 0.08 percent to 0.05 percent for the rest of the UK too and in the US as well! The evidence of the impact of this change in Scotland hasn't yet come through. But according to a recent report, other countries that have reduced their drunk-driving limit by the same amount have seen significant falls in road accidents and fatalities.[12]

It is estimated that lowering the BAC in every US state to 0.05 percent could potentially save up to 1,790 lives a year. In 2015, the National Highway Traffic Safety Administration published a report that found that the total economic cost of motor vehicle crashes in 2010 was $242 billion, or 1.6 percent of America's GDP.

Reducing the limit to 0.05 percent would reduce the accident risk by half, even though it wouldn't halve the amount people can drink. At 0.05 percent the risk of an accident for the average person is only slightly increased from having nothing to drink. Proportionally, society would gain a lot more in terms of reducing risk than we'd lose in terms of drinking.

Again, though, for young drivers even this lower amount isn't safe. Some countries insist that learners and those who have just passed their driving tests avoid alcohol completely; they risk losing their license if they test positive for any alcohol.

There has been some preliminary data from Scotland that's not in-line with what has happened when other European countries lowered the legal limit. A study by the University of Glasgow, published in *The Lancet* in 2018, compared the rate of accidents in England and Wales with those in Scotland, and there was no reduction, according to Jim Lewsey, Professor of Medical Statistics at Glasgow University's Institute of Health and Wellbeing. He said the reduction in the drunk-driving limit "simply did not have the intended effect of reducing RTAs." He also said this went against most other international evidence—and that the most likely explanation is that the changes weren't backed up by enforcement by police.[13]

I'd add that at the moment, in order to Breathalyze in the UK, the police need to suspect that you are drunk. But in other countries the police can do random testing, which probably acts as a more powerful deterrent to drinking.

I am also in favor of in-car Breathalyzers, otherwise known as alcohol ignition interlock devices. These are fitted on the cars of convicted drunk drivers in some countries; they don't allow the car to start at all if the driver is over the limit. There is a plan to introduce them in all new cars sold in Europe by 2025. Eventually this approach might significantly reduce drunk-driving accidents, though they are not completely fool-proof.[14]

Another vital way to reduce road accidents from alcohol is to educate people that taking other sedative drugs with alcohol adds to the risks, as shown by the large Europe-wide study, Driving Under the Influence of Drugs, Alcohol and Medicines in Europe (DRUID).[15] This study collected blood from thousands of people who had been involved in traffic accidents and measured their levels of a range of common prescription and recreational drugs. One of its conclusions was that combined consumption is "most dangerous."

You could argue functional testing is fairer, as it shows if you are capable of driving—or not. Although even if you are physically coordinated, your risk-taking faculties might be seriously impaired. The other issue with functional testing is that it doesn't pick up whether you might fall asleep at the wheel.

EIGHT REASONS NOT TO DRINK AND DRIVE

1. It's much harder to concentrate and focus on one task.
2. You need to see—and drinking can give you blurred vision, bad night vision, and/or even double vision.
3. Messages about what you're seeing take longer to reach your brain.
4. You are more confused about what you see, so it takes your brain longer to reach a decision to stop or turn to avoid a crash.
5. Instructions from your brain take longer to reach your body, and you find coordination harder.
6. You lose the ability to judge distances.
7. You are overconfident and also more likely to take risks.
8. You may feel more sleepy and be more likely to fall asleep.

THE DEATH TRAP?

Any amount of alcohol in your system increases your risk of having an accident. And if you do have an accident, the toxicity of alcohol makes it harder for doctors to treat you. It becomes even more dangerous if you have taken any kind of sedative drug such as Valium, as the effect of the two substances add together.

My local hospital—St. Mary's in London—has long campaigned to reduce the harms of alcohol by early intervention, based on their findings that over 15 percent of patients admitted to the resuscitation room of their emergency (A&E) department had a positive blood alcohol concentration over

0.1 percent. If you look at the longest column in the figure below, what it shows is that about one third of people admitted to the resuscitation room and scoring positive for alcohol had blood alcohol levels of 0.24 percent (three times the drunk-driving limit) or more.[16]

Fig. 4. Blood alcohol levels in A&E patients

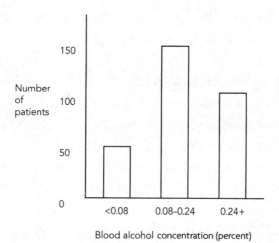

Source: www.ncbi.nlm.nih.gov/pubmed/20497803.

DRIVING THE MORNING AFTER

Did you know you can still be drunk after you've been to bed and woken up the next day? In my experience, very often people don't think about the fact that alcohol takes a while to process.

You don't have to have drunk an enormous amount to still be illegal the next day. The calculator at www.morning-after .org.uk shows you how long you should wait for all the alcohol to leave your system.

If you drink two 5-ounce glasses of 13 percent wine, for example, you won't be allowed to drive (legally) for four and a half hours.

And if you drink three 5-ounce glasses of 13 percent wine, you'll have to wait eight hours to be under the limit.

That said, even if you are legal to drive the next day, you may not be safe. A review by the University of Bath in the UK was very clear on this: numerous studies show that being hungover affects key skills you need for driving, including coordination, memory, and attention.[17]

WOULD A HIGHER LEGAL DRINKING AGE SAVE LIVES?

Probably the most remarkable public health intervention in terms of reducing deaths on the road happened in the US in the 1980s. It was the increase in the drinking age, from 18, 19, or 20—it varied by state—up to the age of 21. The US Department of Transportation forced each state to do this by threatening that if they didn't comply, they would lose 10 percent of the federal highway building budget (a lot of money).

It is controversial, but it's been claimed that this law—the National Minimum Drinking Age Act—may have saved hundreds of thousands of lives. Young people's driving is much more impaired by a given level of alcohol, and in the vast majority of states this new law has massively reduced drunk-driving deaths on the roads among under-21s.[18]

Despite this remarkable data from the US, I'm not sure I think raising the minimum age of drinking is a good idea. If someone can vote, if they can join the army, fight, and die, get married, then they should be allowed to buy alcohol.

ALCOHOL AND VIOLENCE

Fig. 5. Alcohol and violence

Serious violence
1 in 2

Traffic offenses
1 in 4

Traffic crashes
1 in 5

Drug and antisocial offenses
1 in 2

Family violence
1 in 3

Property offenses
1 in 4

Sexual offending
1 in 5 [1 in 2 not known]

Violence
1 in 3

18 percent of police budget

Source: New Zealand Police 2010; www.alcohol.org.nz/research-resources/nz-statis
tics/alcohol-and-crime. Accessed October 2014.

Alcohol makes people violent in the same way that gasoline makes a fire worse. Children fight because they don't have the words to sort things out. You'd imagine adults would have grown out of that. But human beings—unlike most animals—don't need a reason to attack our own species. At least not when alcohol is part of the picture.

> **AUTHOR ALEKSANDR SOLZHENITSYN DESCRIBES IN *THE GULAG ARCHIPELAGO*** how, during the Second World War, the Russian army would ply disposable troops with drink before an advance (these were the men whose guns were made of wood).

Once drunk, they'd be told to walk toward the German lines. The idea was that the Germans would use up their machine gun bullets on them, and once that had happened, the Russian troops with real guns would be sent in to kill. The men who were ammo fodder probably wouldn't have been able to do this if they hadn't been drunk.[19]

Much further back in history, the Vikings got high on a combination of alcohol in the form of mead and magic mushrooms before their berserker raids. This was supposed to give them magical powers. It would definitely have made them seem immune to pain.[20] One theory says that the binge drinking culture found in modern-day northern Europe reflects the spread of Viking genes throughout this region.[21]

A WHO report states that in 2016, violence where alcohol is involved caused 90,000 deaths globally a year.[22] The US Bureau of Justice Statistics found in a survey of inmates in state and federal prisons that "an estimated 37% of state prisoners serving time for a violent offense in 2004 said they were under the influence of alcohol at the time of the offense."[23]

What is true is that alcohol makes you—whether you're a man or a woman—more vulnerable to becoming a victim. You're not as aware of the threats as you would be when sober and are less able to defend yourself if you are attacked.

How people transform when drunk—from mild-mannered Jekyll to violent Hyde—can be shocking. Some people who seem perfectly sane, normal, and loving when sober commit the most violent and terrible acts when drunk. An example of this occurred in the UK in 2012 when Andrew Hall, a Stoke City youth team footballer, then age 18, stabbed his 15-year-old girlfriend, Megan-Leigh Peat, over 60 times, which his legal defense described as a "catastrophic loss of control."[24]

There are a few possible explanations as to why people change in this way. People who've had a violent upbringing tend to perpetuate that in later life. At the level of psycho-analysis you could say that the person is unhappy with them-self and that their violence is simply them projecting that unhappiness onto others.

When people black out, they can do things—some extremely violent—that they later profoundly regret. Some people can't even remember what they did.

There was an interesting but horrific case I was involved in. A man stabbed his roommate to death, with a screw-driver in the eyes, 50 to 80 times. He pleaded not guilty to murder. His defense was that he did it during a blackout. His blood alcohol level was extremely high—high enough, in my opinion, to make him black out. But the court decided it was murder. Being in a blackout is not a defense because the person has made the choice to drink.

One way we can think about alcohol-induced violence is to divide it into two types: defensive and aggressive.

1) DEFENSIVE This is violence as an unintentional reflex. It's when someone bumps into you (assuming you're the one who gets violent), spills your drink, or "looks" at you in a certain way. In the UK, we even see this at fancy sporting events such as Ascot or Goodwood, where men dressed up in suit and tie for a day of horseracing erupt into a free-for-all fistfight.[25] And it's one reason sporting venues now use plastic glasses. (As an aside, a champagne bottle is one of the most dangerous weapons of all, as when it hits someone's skull, it won't break.)

One recent study from Bristol University in the UK showed that drunk people misinterpret other people's gazes. Instead of being able to read an expression of, for example, disgust, they see it as anger. This misinterpretation effect was strongest with men rating male faces.[26]

When people feel they're being threatened or having their space invaded, they then overreact to the perceived threat. This happens because alcohol has provoked a failure of the higher control centers of the brain, and when you can't control yourself, you overreact. Being drunk releases a primitive, defensive aggression that you'd normally have intellectual control over. Your frontal cortex is no longer working optimally, so you are unable to put in place your usual rational approach to what you now see as provoking behavior.

2) AGGRESSIVE The classic example of this is a sports fan—or a group of them—planning to riot or fight fans of the opposing team after the match as University of Maryland basketball fans often do after games against Duke, or Vancouver Canucks fans did after the team lost the 2011 Stanley Cup final. This kind of aggression is intentional, done with self-control, and is about letting loose. It's done because the person wants to fight. Drinkers use alcohol for liquid courage and drive, as well as for calories and energy.

AAA: ALCOHOL, ACCIDENTS, AND AGGRESSION

Alcohol works for this because it deadens both fear and pain through its effects on the brain's glutamate and GABA systems, as well as endorphins. Plus the dopamine surge increases energy and drive.

ALCOHOL AND DOMESTIC VIOLENCE

It's an incredibly difficult situation to be in a relationship where one partner—most often a man—becomes violent after drinking.

One UK study looked at domestic abuse statistics in Lancashire (the county home to Manchester and Liverpool) during three World Cup soccer tournaments when England was playing. The number of incidents rose by 38 percent when England lost compared to when they weren't playing. It went up by 26 percent if they won or drew.[27] It's not unlikely alcohol was involved in most of these cases.

Estimates of how many people who commit domestic abuse were drunk at the time range from a quarter to three-quarters. But as studies have shown, the more severe the violence—including rape—the more likely it is that alcohol is involved.[28]

If someone is already violent, they tend to get more violent when they are drunk. If your partner has previously been violent when drunk, it shows they have the capacity and it's likely to happen again if they drink. And that is despite how sorry or regretful they are afterward.

This may sound like old-fashioned advice, but if you are in a relationship like this, try not to have an argument when the violent person is drunk. Don't take on a drunk. They behave in ways they can't control.

Should you leave your partner? Of course you should never stay in a dangerous or violent situation if you can possibly help it. However, I have met men who have managed to stop drinking and remain abstinent, for example, by going to AA, and have not been violent again.

Whether your partner is abusive or not, if you have concerns about their drinking, talk about it with them when they are sober.

And if your partner is abusive, please get help.

For advice, contact the organizations listed in the Resources section at the back of the book (see page 227).

WHAT ALCOHOL DOES TO YOUR MENTAL HEALTH

THERE'S NO DENYING that alcohol makes you feel good—in the short term. Part of the reason for this is that it can numb negative emotions while you are drunk. But it also destabilizes your brain chemistry.

As you withdraw from the effects of alcohol on your brain, you're likely to get psychological symptoms such as anxiety, sadness, lack of sleep, and a short attention span. That's especially true after a night where you've been impulsive or behaved with a lack of judgment, aka drinker's regret. In fact, anxiety the day after drinking is so common, it even has its own nickname—hangxiety or, in Australia, boozanoia. You might get a 4 a.m. wake-up and/or an accompanying panic attack. Repeated withdrawals produce effects on the brain that, long term, add up and change your moods and thoughts, raising your risk of both depression and anxiety.

The relationship between alcohol and mental health is a complex one. Patients often present with both alcohol problems and mental health problems, so it's a bit of a chicken and egg situation: Which comes first? Are you using alcohol as a form of self-medication? Or has alcohol changed your brain and caused or worsened a mental health condition? There's another conundrum: As a doctor, which one should you treat first?

Probably the most common mental health issue linked with drinking is anxiety, followed by depression. But people also drink to push down other difficult feelings, such as the traumatic flashbacks of PTSD or, often, during a manic phase of bipolar disorder. We look at these conditions in more detail in this chapter.

WHAT TREATMENT LOOKS LIKE NOW

It used to be that the treatment of alcoholism and addiction came under the umbrella of mental health services. But now in many countries, including the US, many needy people find themselves on the streets or in prison.

In the UK, we have a real shortage of treatment places and training facilities for health-care providers in addiction. The most disadvantaged group of patients are those with complex mixtures of addiction and mental health issues, who are now largely looked after—if you can call it that—in hospital emergency departments or in prisons. One 2011 study called the US state prison system the "nation's largest mental health institution."[1] The recent massive rise in deaths from opioids is likely due to this change in policy and the resulting reduction in access to treatment. I am firmly of the belief that alcoholism is a disorder of the brain, like other mental health issues, and its treatment should be embedded in psychiatry.

The problem now is that, when people present with both alcohol and mental health issues, they get pushed between mental health services and addiction services, with experts in each saying they should get the other problem treated first. In my clinical practice, I saw this passing of the buck all the time.

Alcoholics Anonymous (AA, more of this later) is an excellent organization, but it shouldn't be the only help available. The reality is, people do best with expert medical and psychological support: medication, therapy, and so on. This is especially true when there are mental health conditions in the mix too.

The evidence as to how alcohol or mental health conditions affect each other isn't clear in the literature. One interesting study from Norway looked at the association between anxiety, depression, and the amount people drank. This was a large-scale population study of nearly 40,000 people.[2] This study found that, when compared with moderate drinkers, nondrinkers had a higher risk of anxiety and depression. It doesn't prove, however, that nondrinkers have more mental health problems; rather, it shows an association. It may be that some people with mental health conditions don't drink as a result. This study also found that heavy drinkers were more likely to have both anxiety and depression.

Conversely, a study from Hong Kong showed that, especially for moderate female drinkers, giving up alcohol will increase your mental well-being.[3] Those who quit alcohol had, after four years of abstinence, levels that approached that of lifetime teetotalers.

These are the mental health conditions that most often occur along with alcoholism.

ANXIETY

It's not surprising social anxiety is common: In evolutionary terms, it is useful. Its biological function is to protect you from strangers who might harm you. Although it's unlikely to feel particularly useful when you're standing outside a party because you can't bear to go in. Most people have some degree of social anxiety, and social anxiety disorder is simply an extreme form of this.

It was while I was working as a consultant at my specialist anxiety clinic in Bristol in the 1990s that I realized how intertwined anxiety disorders and alcohol are.

I was called to do a home visit for a patient who was agoraphobic and so too anxious to come into the clinic. He was in his 40s, but he looked 80. He had peripheral neuropathy—no feeling below his knees—due to his drinking.

We started to chat about his problems. He told me that in order to go outside the house, even just to cut the grass, he needed to drink two cans of Special Brew, a very cheap, potent beer.

I asked him, have you told anyone about your alcoholism? "Yes," he said, "I told my doctor. And my doctor said I needed to go to AA. So I went to AA, but I needed four cans of Special

Brew to go into the meeting. And the truth is, the people there are not like me, because they're not anxious like me."

He was right: you won't be able to access a group therapy program like AA if you can't speak to the other people there.

Looking at the results of one major US study of the co-occurrence of social anxiety disorder and alcoholism, we can estimate that up to a quarter of young alcoholics have social anxiety disorder.[4] In male alcoholics it's the most common condition diagnosed other than alcoholism. And the Anxiety and Depression Association of America has reported that 20 percent of people with social anxiety also have alcoholism.[5] It's a very well-established diagnosis for men, but there isn't as much comparable data for women.

It's also a perfect example of a mental health vicious circle. You start as someone who's very anxious about a social situation. So you drink to get over it, self-medicate. Each time you do this, your tolerance builds. At some point, one day as you go into withdrawal, you begin to feel even more anxious because the alcohol has changed your brain chemistry.

If we scan the brains of people affected in this way, we can see changes to the amygdala, the fear and stress center. These adaptive changes are epigenetic—that is, they are changes at the level of gene expression; fear- and stress-promoting genes are being overexpressed.

A similar process can happen with other kinds of anxiety-related phobias, for example, agoraphobia. The end result? You are addicted to alcohol and are therefore unable to function without it.

There is another patient I remember vividly, when I was a

trainee psychiatrist in Oxford in the 1980s. He was in his late 20s and presented to our anxiety clinic with severe collapsing panic attacks. Each time he had one, he'd think he was having a heart attack and would be rushed to the hospital.

His story was that from the age of 17, he'd been going to the pub. He felt anxious about talking to anyone, but the thought of talking to girls was the worst. So he drank to deal with his shyness. And that was how he socialized, every night. And his brain got tolerant to this regular alcohol intake. Then one night, after ten years of doing this, just as he was going into the pub he had his first massive panic attack.

What had happened was, over the course of the day, he'd gone into alcohol withdrawal. The brain chemicals involved in this made him feel anxious. At the same time, in anticipation of the alcohol he was about to order, his brain was turning off his GABA system, which increased his anxiety levels even more, a sort of neurotransmitter double whammy. So his anxiety levels rose and before he could douse his anxiety with alcohol, it reached a crescendo and he panicked.

The challenge for every doctor when faced with a case of anxiety and alcohol dependence is, do you treat the drinking or the anxiety? The alcohol specialist will say, you've got to stop drinking. But the patient can't stop drinking, as he's too anxious. The anxiety specialist will say, I can't treat you, as you are drinking. So he doesn't get treated. Both therapists blame the alcohol.

My reasoning is this: if we'd stopped him from drinking, he would have gotten more anxious and wouldn't have been able to socialize. What we did—and what we do with people like him now—is treat the anxiety with either a talking therapy (such as cognitive behavioral therapy, or CBT) or an SSRI (selective serotonin reuptake inhibitor) medication. Eventually, this enabled him to cut down his drinking to normal levels.

So how can you tell if your drinking is due to social anxiety? One clue is if you drink before you leave the house to reduce your anxiety about going out.

Another is if you're always first to the bar. A friend who had had alcohol problems, when he was finally sorted and sober, said to me: "You never noticed that I was always in the bar before everyone else arrived. I did that because I knew I had to have two drinks before the rest of you came, or I would be too anxious."

It's worth saying that there are some people with anxiety who cannot bear to drink. Those people—who often suffer from panic attacks—are using all their cognitive capacity to suppress their anxious reactions. They feel that if they drink, they won't be able to do this and will become more anxious.

DEPRESSION

You may have heard that alcohol is a depressant. But do you know what that means? It means that regular, chronic alcohol use affects the serotonin system, disrupting the brain in the direction of having low mood.

Just as with anxiety, a doctor's dilemma when depression and drinking coexist is knowing which is the primary problem.

This might seem like a bit of a pointless question because we know drinking drives down mood and low mood can cause more drinking. But it is clinically important: one of the major predictors for having a poor response to being treated for drinking problems is preexisting low mood. Another issue is that, if you drink while taking antidepressants, they won't work as well.

Alcohol is very often one of the substances that depressed people use to self-medicate. I have often said to patients: If you get depressed after drinking, be very careful. If you feel high when drinking and then low when hungover, be careful. These are both risk factors for becoming alcoholic, as they can be a driver to getting you into a cycle of drinking. If this sounds like you, I recommend being aware of this risk and using the strategies in Chapter 9 to avoid becoming dependent. If the mood problems continue in the absence of drinking, you may need treatment, perhaps a mood stabilizer or antidepressant. See your doctor and discuss it with them.

PTSD

People suffering from PTSD often turn to drink to suppress memories and flashbacks and in order to get to sleep. Just like with anxiety disorders, self-medicating for PTSD with alcohol is a Band-Aid that can lead to alcoholism.[6]

Despite the high profile of the condition, there is still very little research into the most effective means of preventing or treating PTSD. The usual approach of debriefing has been shown to be without effect, and there isn't yet a drug with enough evidence. This may be because most drug treatments are initiated well after the trauma, which may be too late to prevent the laying down of immutable brain traces for the memories, behaviors, and effect that trauma causes and that develop into PTSD. Which leads me to an interesting paradox. Some studies have found that if you are drunk when you are traumatized, you're less likely to get PTSD in the first place. That's because although you may (or may not) be able to remember the facts of what happened, you are too drunk to lay down emotional memories.[7]

SUICIDE

People who have committed suicide often have high blood alcohol levels. And close to half of all people who commit suicide are intoxicated. This could be due to needing liquid courage to go through with it, or it could be that they attempt suicide because they are disinhibited by alcohol.

Also, we know that when people are drunk, they are more likely to attempt suicide and also to use more effective methods. According to WHO's *Global Status Report on Alcohol and Health 2018,* you have a seven-times-increased risk for a suicide attempt soon after drinking alcohol, and this increases to 37 times after heavy use of alcohol.[8]

In addition, being an alcoholic means you're more likely to try to commit suicide. One study found that the most dangerous time of all is at the end of a binge or the start of the withdrawal period.[9]

Alcohol intoxication is also associated with attempted suicide—now called deliberate self-harm (DSH) and discussed later—again through a process of alcohol-induced brain disinhibition and despair. Preventing suicide

> and self-harm is very difficult, but for people with underly-
> ing depression, antidepressant medications can be very
> effective, and for those where the primary driver is alcohol,
> stopping drinking through medication or psychological
> treatments will be helpful.

OTHER MENTAL HEALTH CONDITIONS AND ALCOHOL
ADHD/IMPULSIVITY

People who are very impulsive—they may have been diagnosed with ADHD or an impulse control disorder—may be more vulnerable to abusing alcohol. For example, studies suggest that the likelihood of becoming alcoholic for an adult diagnosed with ADHD is five to ten times higher than in the general population.[10]

ADHD is thought to occur because the top-down control of behavior from the brain's frontal cortex isn't sufficient to control the bottom-up drives and impulses. This failure of control may be because the dopamine and noradrenaline systems in the frontal cortex haven't developed adequately. That's why the most effective treatments for ADHD are medicines that enhance dopamine function (e.g., amphetamines or Ritalin [methylphenidate]) or those that increase noradrenaline activity (e.g., Strattera [atomoxetine]). The other main treatment is CBT. This helps the person strengthen their frontal control regions and also learn ways to limit their urges.

OCD

People diagnosed with Obsessive Compulsive Disorder (OCD)—repeated checking in the face of knowledge—tend to be less likely to use drugs and/or alcohol. But some do use alcohol, in order to decrease the anxiety that exists within OCD. And if they do use it, they are more likely to become addicted.[11] That's because the rituals of OCD become attached to the rituals of drinking. In fact the relationship is so strong that there are addiction experts who believe that repeated alcohol use is for many people—even those who haven't been diagnosed with OCD—more a form of compulsion than one of craving, and a scale to measure this has been developed—the Obsessive Compulsive Drinking Scale.[12]

As with the other psychiatric disorders that occur with alcoholism, treatment requires addressing both problems. For OCD, antidepressant medicines such as SSRIs are the first-line treatment, along with CBT. It is possible that, as with the anxiety disorders, treating the OCD can help reduce the dependence on alcohol.

BULIMIA

Bulimia is an eating disorder that involves binge eating followed by fasting, by making yourself vomit or purge. There's an interesting theory, which started in the 1980s, that bulimia is the female equivalent of alcohol dependency, that they are two sides of the same coin.[13] And it's true that a lot of people who are bulimic abuse both alcohol and drugs. In one study, half of the subjects with bulimia were also alcohol-dependent.[14,15]

SELF-HARM

Drinking increases the risk of self-harm—injuring yourself. A 2005 review of 10,000 cases of self-harm showed that women tended to drink excessively particularly around the time they self-harmed (also see the section on suicide, page 99).[16]

BIPOLAR DISORDER

People often present with both bipolar disorder—having manic and depressive episodes—and alcohol abuse. Some people drink more during the high of manic episodes, some self-medicate for their feelings of anxiety. Studies suggest it may be that one triggers the other.[17]

PARANOIA AND PSYCHOSIS

It's not uncommon for very heavy drinkers to end up developing paranoia or a psychosis-like state that can make them very suspicious of other people. This is due to changes in the brain, caused by heavy drinking over time, that are similar to those seen in people with schizophrenia and that involve the dopamine system. Although rare in the general population, this phenomenon is well known to psychiatric services, as it can present as very extreme and often is mistaken for paranoid schizophrenia. Both illnesses respond to treatment with antipsychotic drugs, but, of course, to keep alcohol psychosis at bay, it's critical to cut down on drinking too.

If any of the previous symptoms sound familiar, don't ignore them. Can you talk about them with your partner or a friend? Before going to see your doctor, keep a diary for a month, including your mood or eating behaviors or obsessive checking—whatever your symptoms are—along with your drinking record. That will maximize the value of your consultation with the doctor and help them work out the best way they can help you. If you have suicidal thoughts, seek help immediately.

HORMONES AND FERTILITY

WE TEND TO use the word hormones as shorthand for things like ovulation, pregnancy, and periods, but as you probably know, they're much more complex than that. The endocrine system is a whole network of brain and nerve cells and secretory glands producing chemical messengers that interact with one another and affect all of the body and the brain. The effects happen well below our consciousness, but we all experience them when we are stressed, angry, in love, hungry, tired, and so on.

Hormones set the background level of most functions in the body. They are the most important homeostatic regulators, keeping our body running smoothly hour to hour and day to day. So if they go wrong, it can lead to real health problems.

For instance, as well as reproductive hormones, there are thyroid hormones, growth hormones, and stress hormones. Each of these hormones can either directly change how our body functions or they can turn on other hormones—or both. All of these hormones don't just come out of the brain and the glands where they're produced, they feed back into them too, switching themselves and one another on and off. It's a system that's always trying to get itself back into balance; when you add in alcohol, it can quickly get out of whack.

Take what happens when you go into withdrawal from alcohol after a heavy night out. The main system where hormones are produced and interact is called the hypothalamic-pituitary-adrenal (HPA) axis. These three glands produce hormones that control the stress response. This happens when your brain or your body gets stressed—which includes when you are coming down from alcohol, aka withdrawal.

There are two parts to the stress response.

1) There's a stress pathway direct from the brain to the adrenal glands, which sit on top of the kidneys. The adrenal glands release the stress hormones adrenaline and noradrenaline. And these are the hormones that make your heart pump faster, make you feel alert and hyperactive, and wake you up at 4 a.m., sweating and nervous.

2) The brain turns on the hypothalamus, which releases a hormone called CRF (corticotrophin-releasing hormone). This travels in the blood down a special set of veins to the pituitary gland. This is the master gland of the endocrine system and controls most hormones. It sits just under your brain, in the middle of the head at the back of your eyes.

When the CRF hits the pituitary gland, it stimulates the release of another hormone called ACTH (adreno-cortico-trophic-hormone, in case you're interested). ACTH travels through the bloodstream to the adrenals and prompts them to release the stress hormone, cortisol.

The noradrenaline/adrenaline and cortisol systems work together to provide a coordinated response to stress. The first part of the stress response happens immediately; then, over the next minutes and hours, cortisol rises. If you drink every day, and thus turn on that system repeatedly every morning as you go into mini withdrawal, you're pushing yourself toward high blood pressure. You always have to have some cortisol in your blood to keep you alive, but if it gets too high, tissue damage occurs. Too much cortisol is like taking steroids: it alters your ability to metabolize sugar, so it also pushes you to diabetes. And, over time, it can also drive changes in the brain that can lead to and perpetuate depression.

WHAT ALCOHOL DOES TO SEX HORMONES

Prolactin is the hormone that promotes milk production. And it's been shown that levels are higher in heavy drinkers of both sexes.

In women, prolactin also has a role in regulating ovulation. Even a single dose of alcohol can increase prolactin levels, although it is dose dependent. One study of women showed that a single episode of drinking moderate to high amounts of alcohol—around four drinks—made prolactin rise and oxytocin (the bonding hormone) go down.[1] The effect is thought to be transitory, but if you drink every day, levels will stay high. And one theory says that high levels of prolactin may be behind the increased risk of breast cancer that comes from drinking.

Men need low levels of prolactin for sperm production. But if levels rise, you can get breast growth and testosterone levels go down. This is one reason men who drink heavily end up being feminized. Alcohol also lowers testosterone via other mechanisms; it increases levels of an enzyme that breaks it down and turns it into estrogen. This can happen even before alcohol starts to cause serious liver problems. And acetaldehyde (produced in the liver when alcohol is broken down) impairs the production of an enzyme that's key for testosterone synthesis.

Because waste hormones are broken down in the liver, once liver disease begins, a man will not be able to metabolize estrogen as well as before. So he will become even more feminized, possibly grow breasts, and lose his facial hair. The feminization is also partly due to a drop in IGF1—insulin derived growth factor—which leads to less muscle, less testosterone, and less sperm. Male alcoholics also lose muscle mass. Besides not eating properly, one of the reasons is that they don't manufacture enough HGH (human growth hormone).

DOES DRINKING REDUCE FERTILITY?

The short answer is, yes. Alcohol causes the production of destructive elements called free radicals, which can damage proteins and DNA. You may have heard of free radicals already, in terms of sunlight causing free radical production in the skin. Although it's normal for your body to produce free radicals, the addition of alcohol means too many are produced and in the wrong place, so they then harm the body. It's thought many of the toxic effects of alcohol in the liver, but also in the ovaries and testes, are due to free radicals. This process is called oxidative stress because the body's proteins— including DNA—are oxidized (damaged).[2]

Experts agree heavy drinking does decrease male fertility. There are obvious ways: It lowers libido and increases the likelihood of impotence too. Then there is its effect on sperm. The jury is out on whether a (very) moderate amount of alcohol decreases or increases sperm quality and volume.[3] One study showed a higher sperm count for moderate drinkers (in this case, around three to five pints of beer a week) than those who drank less.[4]

But another study showed that drinking any more than three drinks a week (2.5 pints of beer) affected sperm quality and quantity. And the more men drank, the bigger the effect.[5,6]

There's a similarly mixed picture for female fertility— although the general message is the same: that alcohol probably does damage fertility. One study showed fewer than ten drinks a week doesn't seem to have a measurable effect on female fertility.[7]

However, a review of studies of nearly 100,000 women showed that drinking at any level reduced fertility—13 percent (for any drinking), 11 percent (for light drinking: around one beer or a very small glass of wine a day), and 23 percent (for moderate-to-heavy drinking: more than a small glass of wine a day).[8]

In another study, 50 percent of social drinkers (people who drink up to four drinks a day) and 60 percent of heavy drinkers (four to eight drinks a day) had disruption in their menstrual cycle and altered sex hormones. Half of the social drinkers who had more than three drinks a day had anovulatory cycles—that is, they didn't ovulate.[9]

CAN I DRINK WHILE I'M PREGNANT?

More than 1 in 9 pregnant women in the US drink alcohol while pregnant, and about 4 percent binge drink, according to new survey estimates from the Centers for Disease Control and Prevention. Estimates of current drinking and binge drinking among pregnant women in 2015–2017 were both slightly higher than in 2011–2013—when the figures were 10.2 percent and 3.1 percent, respectively.

"Drinking alcohol while pregnant can cause miscarriage, stillbirth, and fetal alcohol spectrum disorders," the report says. Alcohol in the mother's blood that passes through the umbilical cord to the fetus can lead to physical issues, behavioral problems, and intellectual disability. "There is no known safe level of alcohol use during pregnancy or when trying to get pregnant," the CDC says.[10]

This is because evidence suggests alcohol increases the risk of stillbirth, premature labor, and fetal growth restriction. Babies of drinking mothers also run the risk of withdrawal straight after they are born—they can even experience the DTs (delirium tremens) if they are not diagnosed and treated.[11]

Drinking also increases the risk of fetal alcohol spectrum disorders (FASD), a range of mental and physical problems caused by exposure to alcohol during pregnancy. As a poison, alcohol crosses the placenta and goes into the developing fetus, potentially causing a huge range of lifelong issues.[12]

The most easily diagnosed type of FASD is called fetal alcohol syndrome, in which the baby has specific and identifiable facial features: a smooth philtrum, thin upper lip, small eyes, low-set ears, small head size. But these children, it's thought, only account for around one in ten children who have FASD. The obvious physical changes are due to the mother drinking during weeks six to nine, when the facial features are forming. But it is the other aspects of FASD—caused by drinking at different times during the pregnancy—that are possibly the biggest burden. These include disturbed and hyperactive behavior, low intelligence, developmental delays, impaired memory, problems with hearing and vision, growth impairments, and problems with the heart, kidneys, and bones.[13]

There's some evidence that bingeing during pregnancy is worse than a more spread-out pattern of drinking. No level of consumption is safe, but the risk seems to go up particularly at around 28 grams—around two standard drinks per week.[14,15]

The global prevalence of FASD is estimated at around 8 in 1,000 in the general population, but it can be much higher in certain high-drinking populations. So it's not a small issue: there are many more affected kids than those with, for example, Down's syndrome, which is 1 per 1,000.[16] The country with the highest rate of FASD is South Africa, where more than one in ten children have it. One reason might be that in some rural areas in South Africa, the vineries pay workers partly in alcohol. This practice was made illegal in 1961 but has been reported to exist still.[17,18,19]

One big issue for parents is that the brain damage part of

FASD will happen during the first trimester. That's why the recommendations are clear not to drink if you are planning to be pregnant. If you discover you're pregnant and you've been drinking, stop. And try not to worry. The reality is, for the vast majority of women, the risk to the child is going to be very low. It's estimated that 1 in every 13 pregnant women who drinks while pregnant will have a child with FASD. And the risk is strongly related to level of consumption.[20]

HOW DOES ALCOHOL AFFECT MENOPAUSE?

It's not clear whether drinking has an effect on the timing of menopause: some studies seem to show it makes it earlier. Others seem to suggest low to moderate drinking might be associated with a later menopause.[21]

What has been shown is that alcohol can make the menopausal symptoms of sweating and hot flushes worse because of its effect on the HPA axis and the resulting increase in noradrenaline.[22]

If you have any of the depression or anxiety symptoms that can go along with menopause, you might find drinking makes them worse too. Menopause makes some women more vulnerable to these because the changes in hormone levels affect neurotransmitters such as serotonin and noradrenaline. These same hormone changes after menopause may alter alcohol sensitivity, so you may find the same amount of alcohol gets you more drunk. Also, both men and women get more sensitive to alcohol as they get older due to a reduction in body fluid, which results in a higher blood alcohol concentration.

Another worry that women often have at this time is the

risk of breast cancer. Your risk goes up with age and, as you already know, alcohol further raises your risk, as does hormone replacement therapy (HRT). In fact, there is likely a cumulative effect on breast cancer risk if you both drink alcohol and do HRT. This means the risk goes up more than if you added the two risks together.[23]

WHY DO I GET MORE DRUNK SOME TIMES OF THE MONTH THAN OTHERS?

Surges and dips in progesterone and estrogen—which happen naturally over the course of your menstrual cycle—change the sensitivity of your brain to alcohol. Progesterone, just like alcohol, works on the GABA system in the brain. Add alcohol to a progesterone surge, and you'll likely feel more drunk.

During pregnancy, it is the super-high levels of progesterone needed to keep the pregnancy that account for memory problems, aka baby brain. High progesterone levels, in effect, sedate you. You could say you are pre-drunk.

Most of the breakdown of progesterone happens in the liver, although it also takes place in the brain, skin, and other areas. Progesterone can be broken down via two pathways. The pathway will depend on the balance of some specific enzymes in your body:

1. In some women, progesterone breaks down to a sedating, anticonvulsive chemical called allopregnanolone. What is really fascinating is that this might help explain postnatal depression. During pregnancy your brain becomes dependent on the high levels of progesterone and adapts to them in a way similar to how your brain becomes tolerant to alcohol.

Then, after birth when progesterone levels drop, you go into withdrawal. Now a new drug that's a synthetic form of allopregnanolone—brexanolone—has just been licensed to treat postnatal depression. It is given intravenously for several days to restore the postnatal deficit.

2. The other breakdown route progesterone can take is to pregnenolone sulphate. The result of this is exactly the opposite to that of allopregnanolone. It decreases the effects of GABA and so causes anxiety and seizures. It feels like alcohol withdrawal. This explains why, as progesterone fluctuates during your cycle, and depending on what else is happening in your body, you may become either more anxious or relaxed.

Your current estrogen levels also influence how alcohol affects you. When estrogen is high, alcohol gets you higher, due to its effect on dopamine receptors. This increased euphoria and pleasure can put you at a higher risk of addiction.[24]

It's likely that the hormonal fluctuation of the perimenopause years are the reason why women find their relationship with alcohol changes, whether it's that they can no longer tolerate it or they want to drink more.

CAN I DRINK AND BREASTFEED?

You may have heard the old wives' tale that stout is good for breastfeeding. Although drinking does increase prolactin, beer's effect on milk production is more likely to be due to the barley extract it contains—which explains why nonalcoholic beer has been shown to work just as well.

113

And if you were wondering how much alcohol gets into breast milk, a study has shown that it is 5–6 percent of the dose the mother receives, a very low amount.[25]

That's why the CDC's advice is that if you are breastfeeding, the limit is the same as if you aren't—ten drinks a week. If you do drink ten drinks in a week, spread them out over a few days. And don't feed for two to three hours after drinking alcohol.[26]

GUT HORMONES AND ALCOHOL DRINKING

The stomach, which doctors used to think of as just a bag that held strong acid and enzymes to digest food, is now known to produce hormones. One of these is called ghrelin (a strange name, but it's based on its role as a *growth hormone-releasing peptide*). Its function is to make you hungry. But it can also make you thirsty; it's been shown to induce craving for alcohol as well as food.[27]

Ten years ago, an influential paper showed that this system determined alcohol liking in mice; when injected into mice, it increased their alcohol intake. And when its effects were blocked, the effects of alcohol in the mice were also blocked.[28]

Since then there have been several studies in humans showing similar effects.[29]

What we've seen so far implies that stopping ghrelin production or blocking its receptor actions would reduce drinking. My research team and I have been studying this in our Medical Research Council–funded study, GHADD, run by Dr. Tony Goldstone, using an anti-ghrelin hormone called desacyl-ghrelin. To check this out, we are brain-imaging abstinent alcoholics while exposing them to drinking cues. Normally these cues activate brain regions involved in craving alcohol, but Goldstone has shown that when desacyl-ghrelin is given, this activation reduces, which we hope may mean it will reduce drinking. Unfortunately desacyl-ghrelin needs to be injected, so it can't easily be used as a treatment. Research is underway, however, to find orally active versions. Once these are available, we plan to test them.

6

HOW ALCOHOL AFFECTS YOUR QUALITY OF LIFE

YEARS BEFORE ANY health issues show up, alcohol is already affecting every organ and system of your body, including your brain. And there are more subtle changes happening too. It's affecting your health, yes, but also your wellness: your performance at work, your skin, your fitness, your sex life.

In this chapter, I hope to answer the questions: Is how much I drink adding to my life, or taking away from it? What's the level of drinking that will make me feel good but allow me to function well too?

ALCOHOL AND SLEEP

After a night out, do you tend to suddenly wake up in the early hours, feeling alert and awake? What you are experiencing is a mini withdrawal from alcohol. This hyperactive, fight-or-flight state is prompted by a surge of the neurotransmitter noradrenaline. You may also be sweating, and you may find it hard to get back to sleep.

What can you do to stop this? You'll need to work out, through trial and error, the limit that affects your sleep. Research suggests you'd have to drink two drinks or more to disrupt your sleep significantly. But first, two drinks is only a couple of beers or a glass of wine. And second, your limit will be personal to you.

If you keep drinking enough to go into withdrawal every morning, the stress response will accumulate. Then, as well as the spike in noradrenaline, your cortisol levels will go up. This alters your ability to metabolize sugar and can push you toward diabetes or a rise in blood pressure. Cortisol may also drive changes in the brain that underlie depression.

Some people get into a habit of drinking to fall asleep. And it works: alcohol puts you to sleep faster and into a deeper sleep more quickly too. It wasn't uncommon, when I was first working on a hospital ward in the 1970s, for doctors to prescribe a glass of sweet sherry to elderly patients at bedtime.

The problem is, in the second part of the night your sleep becomes more disturbed. You may not get as much REM (rapid eye movement) sleep, which is when you dream and the brain consolidates memories.

If you are using alcohol to sleep, be careful: you're increasing your risk of dependence. The best way to improve your sleep is to improve your sleep hygiene. For example, exercise during the day, in the morning and ideally outside to get sunlight, both of which help reset body sleep-wake rhythms.

Wind down for bed without any screens, and sleep in a cool, dark room. You could also try cognitive behavioral therapy; there are some good introductions to it online, such as the Sleepio app.[1] This may be covered under your health insurance; even if you have to pay, the cost will probably be offset by your savings on alcohol.

If neither of these strategies work, do see your doctor. A short course of sleeping aid might help you reset your sleep rhythm. If the insomnia persists, ask your doctor for a referral to a sleep expert.

Do you snore? If so, you've probably been told that you're much noisier after a drink. Snoring is a sign of the beginning of an obstruction in the air going in and out of your lungs. Drinking makes it worse because alcohol relaxes your muscles, including those of your pharynx and palate, the soft tissues at the back of your mouth and in your throat.

Snoring can develop into sleep apnea: this is where your airways collapse so you can't get any air into or out of your lungs. It deprives your brain of oxygen, a condition called hypoxia, just as if you had a plastic bag over your head.

A hypoxic episode can last 20 or 30 seconds until your brain gets so stressed that it fires off a massive fight-or-flight response. You then take an abrupt breath and your airways open for a few minutes, until you go back into a deeper level of sleep. Usually the stress response doesn't wake up the person having it, but the snoring and spluttering that accompany it can be very irritating—and possibly worrying—to any bed partner, who may feel that the apneic person is never going to start breathing again. In my sleep clinic I would come across nondrinking partners staying awake for hours "just in case" they had to revive their partner.

For many people, this pattern of hypoxia followed by a mini stress response repeats itself, sometimes hundreds of times a night. Over days and weeks, each episode adds up, which can lead to daytime sleepiness but also problems such as hypertension and obesity.

Treatment involves keeping the airways open with a Continuous Positive Airway Pressure machine (CPAP). The downside of these machines is their noise; they tend to keep the person's poor partner awake and may be even worse than the snoring.

Both snoring and sleep apnea are more likely if you're overweight. There's a vicious circle I used to see quite often in my sleep clinic: fat men who snore and drink and keep getting fatter. Poor sleep increases ghrelin, the hormone from the stomach that increases your appetite for food and for alcohol too. It also reduces the level of the satiety hormone leptin. (Spikes in ghrelin may be partly responsible for the fact that you feel so hungry the day after drinking too.[2])

A woman in her mid-20s once came in to the sleep clinic with terrible alcohol dependency. She was drinking half a bottle of vodka every night and suffering withdrawal in the daytime. I asked her why she only drank at night and it turned out she was so anxious about going to sleep, she couldn't go to bed unless she was drunk.

The reason for her anxiety was because she suffered from sleep paralysis. This is when your brain wakes up, but your body doesn't, so you can't move and you feel as if you can't breathe. She told me it felt like someone sitting on her chest. Another patient described trying to get out of this state of paralysis as being as exhausting as "climbing up the curtains using only my arms." You can see why, with that kind of unpleasantness in store every night, someone might turn to alcohol to help them get to sleep.

The woman told me that she'd wake up two or three times a night, feeling as if she was dying. The irony was, although the alcohol got her to sleep, the withdrawal was making her episodes of sleep paralysis more frequent. We treated her by helping her to stop drinking, along with a short course of a benzodiazepine sleeping medicine while the SSRI medicine we also prescribed got to work. SSRIs are usually described as antidepressants, but they are very good for anxiety disorders too, and I have found them useful in a range of different anxiety-related sleep disorders including sleep paralysis and night terrors.

Night terrors are different from nightmares: terrors come during the very deepest stages of sleep, unlike nightmares, which come during dreaming sleep. People will often scream and shout, might even get out of bed to fight off whatever they feel is attacking or threatening them. They appear terrified, but they are deeply asleep. Terrors are relatively common in children, who usually grow out of them, but adults get them too.

Alcohol makes night terrors more likely because it pushes you into deep sleep. Night terrors are due to some parts of the brain being awake from fear while the consciousness areas are still asleep. And alcohol helps keep them unconscious.

I once treated a 21-year-old student who came to my sleep clinic after a holiday in Europe with a university friend and his parents. When the vacation rental was full of guests, the two students were asked to spend a night in a tent outside, in the woods. After an evening socializing over food and wine, my patient went to bed. In the middle of the night, during a night terror, he thought he was being attacked by a bear. He felt its paws on his throat so, to defend himself, he bit it. But what had really happened was, he'd been screaming, his friend had tried to wake him by putting his hands on his shoulder—and he'd almost bitten off his friend's thumb!

Like most patients with night terrors, his were worse when sleeping in new places. We often had university students coming to our clinic in their first few weeks at college when they were away from home for the first time, and drinking more too.

But the strangest sleep problem linked to alcohol I've come across was two sisters who had Sleeping Beauty syndrome, aka Kleine–Levin Syndrome (KLS). This is a rare neurological disorder where people—usually teenagers or young adults—fall asleep for three or four days at a time. We don't know what causes KLS. We think it comes from some form of inflammation—perhaps a virus—damaging the wakefulness-promoting centers of the brain. There's almost certainly a genetic component as well.

The same thing happened to each of the sisters, albeit three years apart: The very first time they drank alcohol, they fell into a deep sleep for days. Afterward, they rarely drank again, and never to get drunk. We know from hangover research that withdrawal can lead to considerable inflammation of the brain (see Chapter 2), so perhaps this was the case for them.

ALCOHOL AND YOUR APPEARANCE

Alcohol dehydrates your whole body, but your skin is where you'll see it. In fact half a standard drink makes you pee out roughly 80 milliliters extra of urine. So if you drink one 5-ounce glass of 13.5 percent wine, you'll pee out nearly 200 milliliters extra. Which is why it's a good idea to hydrate before drinking, have a glass of water with every glass of wine, and a pint of water before bed too.

Dehydration not only makes your skin look less plump but also makes dark eye bags more visible because it dilates blood vessels (vasodilation). This is also why drinking often leads to spider veins and to rosacea, which looks like flushing and redness on the cheeks. In one study, the more women drank, the higher the risk of developing rosacea. The worst kinds of booze for this? White wine and spirits. Nota bene: in this study, red wine made existing rosacea worse but didn't provoke it.[3]

The body-wide inflammation provoked by alcohol will show up on your skin as puffiness. And inflammation is likely to be why some people find alcohol is a trigger for their psoriasis too.

In terms of aging, alcohol is bad for all the organs of the body, from the heart to the liver to the brain (see Chapter 2). So it's not surprising that, when a Danish study looked at heavy drinkers, they found that they had more outer indicators of inner aging than people who drank less.[4] Is cutting down good for your skin? According to a survey by the University of Sussex, half of the people who quit for Dry January said their skin improved.[5]

ALCOHOL AND NUTRITION

When Nigel Lawson, a high-ranking British politician, was asked how he managed to lose a lot of weight in a short time, his answer was that he stopped drinking. He explains why brilliantly: "When I've had half a bottle, I forget the reason for not drinking the second half. Or a second bottle. Along with the reasons for not ordering the prawn cocktail, the roast beef, Yorkshire pudding and the apple crumble. So Chardonnay has to go, in my experience, not just for its calories (700 a bottle) but for its unintended consequences."[6]

He's right. Alcohol takes away the higher functions of the brain and makes it more likely that you will eat uninhibitedly. (But you probably already know that.)

Let's not forget about the calories in booze too—it is made by fermenting sugar or starch, after all. One estimate is that alcohol can make up as much as 16 percent of adult female drinkers' total energy intake.

Alcohol calories contain minimal nutrition. I have often seen recommendations for people who want to lose weight to drink spirits with soda. But pure spirits are likely not the best regular drink in terms of health. In theory, one standard drink of any alcohol will have the same toxicity for the liver and other organs, but both beer and wine may possibly contain chemicals with some health benefits. You'll have heard of the claim that substances called polyphenols in red wine are good for your heart (see "The French Paradox," page 62), as well as being protective against cancer and possibly antiaging too. And there's data to show that beer drinkers don't seem to die as young as spirits drinkers, possibly because of the amino acids and peptides beer contains.[7]

In terms of energy, beer is carbohydrate rich and contains around 180 calories a large glass. But studies suggest it isn't specifically responsible for the "beer belly."[8]

The typical hard, sticking-out domed gut is made up of visceral fat—fat between the organs—and probably some of the soft and flabby kind of fat—subcutaneous fat—too. Both are made by excess calories rather than beer itself (although, see the hormones section for the possible estrogen-like effects of alcohol).

Drinking can also feed into a cycle of weight gain, as explained in the sleep section previously. If you drink too much and don't get enough good-quality sleep, it boosts your levels of ghrelin, the hormone that increases alcohol- and food-related activity in the brain—that is, makes you want to eat and drink more. Then you're more likely to eat junk food and, it's been shown, to drink over the guidelines. Which then affects your sleep . . . and so on.

Doctors do wonder why some people get ill from drinking, for example, developing cirrhosis, and others seem to be more resilient. A review in *The Lancet* has pointed out that the health harms of alcohol fall more on drinkers of lower economic status. What this means is, if you are poor and drink too much, you are more likely to get sick. And that's despite the fact that we know consumption of alcohol across the social strata in this country is pretty similar. In the analysis, the researchers show this isn't due to either smoking or obesity.[9]

Diet could be one possible contributory reason. There are obviously other problems that come with being poor: stress, bad housing, not being able to afford heating, and so on. But the real downward spiral often happens for people who drink excessively when they can't afford food. Perhaps a balanced diet provides some protection against health harms? What we have known for a long time is that giving supplements of the B vitamins thiamine (B1) and B12 can (partially) protect the brains of people who are vulnerable by reasons of poor diet. Indeed, as we saw in Chapter 2, thiamine deficiency can lead to permanent brain damage and memory loss in Korsakoff's syndrome. It's thought that this deficiency has a generally negative impact on brains that are already subject to assault by alcohol, acetaldehyde, and inflammation.

If you drink regularly—and have a less than perfect diet—I'd suggest you take a multivitamin supplement that includes vitamins D and K. Or at the very least, take a B vitamin complex.

> **DRINK / CALORIES**
>
> Large glass of regular beer—180 calories
>
> 12-ounce can of regular 5 percent beer—150 calories
>
> Medium glass (5 ounces) of 13 percent wine—130 calories
>
> Glass of champagne—110 calories
>
> Shot of spirits (1.5 ounces)—110 calories
>
> *Source:* Drinkaware.co.uk.

ALCOHOL AND WORK

It seems incredible now that it used to be normal—almost part of the job description—to get drunk at work, to have convivial lunches and schmooze clients. Drinking even took place in consultants' restaurants in hospitals and in senior common rooms in universities. My life as a junior hospital doctor in the 1970s was spent either on the wards or in the hospital bar. Every hospital had one consultant who would regularly be unfit for action because of being drunk. In one of my first senior resident jobs in a top London hospital, I once found a senior professor collapsed on the floor. As I tried to get him up and into an office, the professor of medicine walked by, saw him lying there, sighed, "Oh no—not again!" and helped me carry the incapacitated prof to his office to sleep it off.

During my first stint on a Medical Research Council grant panel in the 1990s, they served wine at lunch. Though not everyone drank, I suspected that grant scoring was more lenient in the afternoon—unfairly for the morning grants. After a few of us pointed this out, the alcohol was discreetly removed from the menu.

Long lunches still happen in some industries, but they are for high days and holidays celebrations and vacations, not the everyday norm. Now most industries are dry, at least during the daytime, and in hospitals drinking is a dismissible offense. The reason for this is obvious—alcohol impairs performance. But also, lunchtime drinking has become unfashionable, so there's social pressure against it too.

In Japan, there's still a tradition that you seal a business deal with alcohol, and drinking after work with your colleagues is seen as obligatory. You must never leave the bar before your boss, which is why workers stay overnight in tiny hotel rooms built just for this purpose.

In China, which now consumes half of the trillion-dollar-a-year global alcohol market, celebratory drinking at meals with the shout of *"Kampai!"* is standard (or at least it seems to be when I am there). I find avoiding getting drunk at every meal the most challenging aspect of being in China, even more so than the language barrier.

In the US as in many countries, companies now use booze as a bonding mechanism, with Friday happy hours and team drinks and, of course, the classic Christmas party. The fallout from those has probably led to more Dry Januaries than any other factor.

It does seem that companies are going to have to become more creative and less alcohol focused with their after-work activities since the advent of #MeToo and the fact that fewer people now drink. That includes former alcoholics as well as those who've chosen to go sober and people who don't drink for religious reasons, such as Muslims and Jehovah's Witnesses.

According to one US survey, one in three people (35 percent) preferred not to drink alcohol at work events.[10] Sixteen percent said they drank at them despite not wanting to, and 12 percent pretended to drink in order to fit in. The biggest morning-after regrets of those who did drink? Revealing personal secrets to colleagues (14 percent) and complaining about work-related issues to the boss or coworkers (13 percent). Nine percent said they embarrassed themselves by getting too drunk and 8 percent said they "engaged in sexual activity with a coworker."

Which brings us to the morning after. Have you ever spent a day slumped at your desk and achieved barely anything? People reckon they are around 40 percent less productive when hungover. *The Atlantic* reports, "Excessive drinking costs the US economy more than $220 billion—or about $1.90 per drink, according to the Centers for Disease Control and Prevention, which studies the negative externalities of alcohol consumption each decade. Seventy-two percent of the costs came from lost workplace productivity, according to the 2006 survey, which suggests that the economic drag from hangovers is about $160 billion."[11]

A review of 19 studies by the University of Bath has shown that it's your sustained attention—concentration and focus—that's hardest hit by a hangover.[12] Your memory is also affected, both short- and long-term, but the problem is in making memories rather than retrieving them.

It sounds obvious, but hangover symptoms are dose dependent. Australian researchers—who Breathalyzed subjects, then tested their memories—found those whose brain function was the most impaired had had the highest blood alcohol concentrations the previous night too.[13]

In short, if you've got a hangover, you won't be doing your best work.[14] So you might as well spend the day in bed (though it may be hard to get a doctor's sick note)!

WHAT ALCOHOL DOES TO FITNESS

You already know that moving your body is one of the best things you can do for your health: it's good for energy levels and for mental and physical health, from your brain to your heart, lungs, muscles, and bones, and it protects against chronic illnesses.

What you may not know is that there's a positive relationship between alcohol and exercise: that is, exercisers are more likely to be moderate drinkers and vice versa. What's not known is why: Is it due to team sports, or a go-hard-or-go-home living ethos? Or maybe some people exercise because they feel guilty about drinking?

However, that doesn't mean alcohol is good for your fitness performance. Not only because of dehydration, which is more severe when you add sweating into the mix. But also because alcohol leads to fluctuations in blood sugar, a drop in muscle strength, and, of course, being tired. It can also disrupt hormone levels—including the growth hormone, which you need for muscle gains.[15,16]

Despite the fact that alcohol isn't great for fitness, it's

surprising how often you come across the two together. Sports teams often use alcohol sales to underpin their finances, and a lot of people enjoy drinking after a game or match.

Drinking in sports has claimed some high-profile casualties, notably the great Olympic swimming champion Michael Phelps. He has been caught driving under the influence of alcohol on several well-publicized occasions and has spoken publicly about his problems and engagement in therapy.[17]

Players on team sports are perhaps more vulnerable due to the long tradition of post-match drinking parties. Baseball legends such as Babe Ruth and Mickey Mantle had well-documented excess alcohol use.[18] Football legend Brett Favre has also recently spoken about his alcohol dependence and binge drinking.[19]

When all professional sports took place on a Saturday, drinking after matches and at weekend parties could often remain hidden. But now that sports are almost a 24/7 activity, it's more difficult to hide your hangover from your coach, so maybe we shall see less of this sports team post-match bingeing.

ALCOHOL AND SEX

Shakespeare had the measure of alcohol and sex. In *Macbeth*, Macduff says: "It provokes the desire, but it takes away the performance; therefore much drink may be said to be an equivocator with lechery: it makes him, and it mars him; it sets him on, and it takes him off."[20]

Because a lot of sex is in the mind, it might be the thought of alcohol being an aphrodisiac that is the real effect rather than any physiological effect.[21]

In one study, 12 heterosexual men and 12 heterosexual women were interviewed about taking marijuana and/or drinking before sex. It found that although alcohol made people feel more attractive and helped them meet a partner, it was also more likely than marijuana to make people choose "atypical" partners and to have "post-sex regret." Both drugs increased the risk of unsafe sex. Unsurprisingly, teenagers are more likely to have unprotected sex when drunk too.[22]

Then there is what alcohol does to sexual function. It's been shown to help women feel more comfortable about looking at sexual images—less inhibited, said the researchers—but it didn't make them feel more aroused.[23]

Women with alcohol problems are more likely to have sexual disorders, such as low libido and the inability to orgasm, although it's not clear if this effect happens with moderate drinking.[24]

As the expression "whiskey dick" (much less charming than the UK equivalent, "brewer's droop") implies, alcohol and erections are opposing forces. The higher the alcohol intake, the more likely a man will have an unsatisfying orgasm, low libido, and/or premature ejaculation.[25]

In the five years of an Australian study that followed 810 men ages 35 to 80, 31 percent developed some form of erectile dysfunction.[26]

As well as alcohol, other risk factors for erectile dysfunction are age, being overweight, and not sleeping enough. And be aware: it's also often an early marker for cardiovascular disease. The good news is, a third of those men were able to improve their sex lives—by cutting down on alcohol, but also by stopping smoking, eating better, and exercising.

But there are some positives to drinking too . . .

DRINKING MAY MAKE YOU MORE CREATIVE

Hard drinking has always been associated with artists, writers, and thinkers, from Jackson Pollock and Beethoven to Dorothy Parker and, further back, Socrates and the ancient Greeks. It was reported by Herodotus, the Greek historian, that the ancient Persians would only finally make a decision after the issue at hand had been discussed both sober and drunk. The idea was, alcohol liberates your thinking—but you shouldn't make a decision when you are drunk.[27]

Conversely, there is some evidence to say alcohol can help you be creative. The psychological theory behind this is, alcohol stops you from focusing well and cuts down your working memory. And so it widens the scope of things you have in mind, and, by doing this, allows you to think out of the box.

In a study done to test this at the University of Illinois in Chicago, the participants were given vodka and cranberry juice, then asked to do a creative word game. It was the kind where you have to think of a word that links three other words (for example, skate, cube, and pack—the link is ice). Compared to those who did it sober, the drinkers were more accurate and faster—and they said their problem-solving felt more intuitive too.[28,29]

DRINKING AND THE NAVY

If you can keep people drunk, you can get them to do what you want. While not exactly a positive in the sense of the rest of this section, you can see how alcohol could be a useful tool in the military. In the seventeenth century, the British Royal Navy started using alcohol as a way to control their sailors. It wasn't a desirable job, so many of the men were press-ganged (forced into service). There's even a story of one poor soul being kidnapped from the steps of the church on his wedding day. Rum rations were there to make the sailors' lives feel less tough as well as give them the necessary courage to scale the rigging when the ship was moving beneath them. And also, rum made them alcohol dependent. That meant they could be controlled by the threat of having it taken away.[30]

In the early 1700s, each naval man over the age of 18 got half a pint of over 50 percent pure rum each day. The sailors' way to prove it was of that strength was by seeing if gunpowder soaked in rum would still burn (like we might set fire to brandy or sambuca today). This is where the term "percent proof" came from. As a pint taken all at once made most sailors too drunk, in 1740 Admiral Edward Vernon split the ration into twice a day and decreed that it must be diluted four times with water. His nickname was "Old Grog," which is where we get the word grog as slang for alcohol. Later on, in an attempt to prevent scurvy, lime or lemon juice was added to the ration.

Even diluted, half a pint of rum is still 16 shots, which is undeniably a lot per day. So in 1823, the ration was cut in half, and then again in 1850. This "tot" of rum (an eighth of a pint measure) given at noon stayed until 1970, when it was replaced by a beer ration, presumably as this was thought to be less impairing.

The US navy too initially allowed drinking, as from 1794 the US congress authorized a daily alcohol ration for sailors. Over time, the regulations of alcohol consumption aboard Navy vessels changed until it was abolished in 1914. However, at the captain's discretion a "beer day" can be held when a vessel has been at sea for 45 continuous days and has more than 5 days left before coming into port. This event is authorized once every 45 days at sea, so once a beer day is held, the ship's crew has to wait another 45 continuous days without a port call until another beer day.[31]

DRINKING AND AIM

Alcohol is a very good treatment for the shakes. That doesn't mean the shakes from withdrawal but just the normal kind of tremor we get when stressed or anxious. That's one reason darts players drink. Champion player Andy Fordham says he was never sober until he went teetotal in 2007. And he says he won the 2004 world title after most of a bottle of brandy and two dozen bottles of beer. "For me, it helped the concentration, numbed everything, you weren't aware of what was going on behind you, you could just concentrate on what was in front of you, the board. I know it was terrible for my health, but it just seemed right."[32]

There's also a condition where you shake, called benign

essential tremor. One of the treatments is alcohol. There was an interesting court case in 2014 of an anesthetist in France who killed a patient by intubating her esophagus instead of her windpipe.[33,34] She had been drinking rosé wine. Part of her defense was that she had to drink in order to stop the shakes. But as she was three times over the drunk-driving limit, she may have been drinking to stop the shakes of withdrawal rather than those of essential tremor. Possibly she had started out drinking to treat her benign essential tremor. . . . This would make her another example of alcohol initially alleviating symptoms such as anxiety or sleeplessness, for example, but then over time leading to dependence.

DRINKING AND CONFIDENCE

There's long been a theory that extroverts are more affected by alcohol than introverts. This is based on work in the 1950s by psychologist Hans Eysenck. The theory was tested in 1958, in one of the very first scientific studies that looked at the relationship between blood alcohol levels and driving performance. It found that impairments to driving started at 0.02 percent and got worse.[35]

In addition, the study found that extroverts who drank drove less well, but introverts were much less affected. In fact, the driving of 30 percent of the introverts actually improved.

This feeds into what we now know about the brain. My theory is, in introverts the GABA system is set too low, so that they receive an instantaneous boost in confidence from drinking. As with social anxiety, if drinking does this to you, be careful: you are more at risk of drinking too much as the incentive to keep using it is stronger.

The confidence effect is also thought to be behind a study that showed people learning to speak Dutch became better after consuming the equivalent of a pint of 5 percent beer; not only did they speak more fluently, but their pronunciation was better too.[36]

ADDICTION: HAVE I GOT AN ALCOHOL PROBLEM?

I drank to drown my sorrows, but the damned things learned how to swim.

Frida Kahlo

IT'S NOT YOUR fault you want to drink. Alcohol is a powerful and addictive drug that's not only legal but enmeshed in our everyday life. It's almost impossible not to come across it as we grow up. And most social occasions and celebrations involve drinking.

Being both relaxing and pleasant to take, alcohol encourages habitual use. Alcoholism is common, even (in fact, especially) in the medical community. There's a joke that goes: What's the definition of an alcoholic? Someone who drinks more than their doctor!

Alcohol can make you feel like the person you want to be too. So when people say, "Oh I adore the taste of my 1984 claret," I find it funny. Because I know that if you drank it for a few days and it didn't get you feeling relaxed and cheerful, you wouldn't be so keen on drinking it.

In the US it is estimated that about 20 percent of the adult population drink above the recommended levels, which puts over 40 million people at increased risk of health harms.[1]

Most people drink to relax rather than to get high and so don't understand addiction. One of the reasons we as a society have to think carefully about addiction is that there's a view, popular with conservative politicians, that addiction is fun and addicts enjoy getting drunk, and that they are weak. What I've seen, for most people who are alcoholics, is that they don't want to do it; they have to do it, so much so that they can't not do it.

There are lots of factors that add up to why people become addicted, which I'll cover in this chapter, including genetics, lifestyle, life events—and the fact that it is an extremely powerful drug.

You could think about it like this: The majority of people who take heroin use it occasionally for fun. But 40 percent of those who try it get addicted and then cannot *not* use it. For alcohol, this figure is around 10 to 15 percent.[2]

Countless addicts have told me they don't want to drink, can't stop, and don't know why. I've seen people ruin their health and their life.

One of the first times the strength of addiction began to be revealed to me was when I was 26. I'd finished my medical training in Britain and, after traveling through India and

Myanmar, I ended up in Australia doing some postgraduate training, and working every weekend as a substitute for a GP. Half of my cases were alcohol related, whether it was people diving into empty swimming pools, self-harming, or having fights.

One morning, around 9 a.m., a woman called me in to see her husband. When I arrived, he was staggering around, really drunk, very confused. He kept saying, "I can't see" and "I don't know what's going on." I ascertained that he'd been out drinking most of the night and had gotten home at 4 a.m. He'd fallen asleep on the sofa and, when he woke up, began staggering around with blurred vision. His wife was extremely angry with him.

I thought, *is he still drunk?* To all appearances, he was. But when I examined him, I found significant inflammation at the back of one of his eyes. He wasn't just drunk. At some point in the night, he must have banged his head; he had a bleed on the brain that was causing pressure on his optic nerves.

At this point, in through the back door came two of his friends, with bottles of brandy in their hands, asking, "Is Steve coming out?" His wife started shouting at them: "No way, you're killing him." Even with his brain injury, he wanted to go with them.

In this chapter, I'm going to explain why it's not people's fault when they become addicted to alcohol, how this happens, and what you can do if this happens to you or a loved one.

WHAT IS AN ALCOHOLIC?

There are a number of ways to categorize drinking levels and the harms they produce. The particularly useful guidelines from the UK Department of Health divides people and drinking into five categories. I'd say that all of category five—Dependent—are alcoholics, and at least some of category four are too.

1. Nondrinker.
2. Low-risk. Regularly consuming 10 drinks or less.
3. Hazardous (increasing-risk drinking). A pattern of alcohol consumption that increases someone's risk of harm. Regularly drinking more than 10 drinks a week but less than 20 for women or less than 30 for men. These people are probably not dependent physically, although they may be psychologically.
4. Harmful drinking (high-risk drinking). A pattern of alcohol consumption that's causing mental or physical damage. Drinking 20 drinks or more a week for women, 30 or more for men. A large proportion of alcohol consumption is done by people who drink very heavily—that is, over six drinks a day. Not all of these people will be dependent, but many will.
5. Dependent. Typically has a strong desire to drink and difficulties controlling its use. Continues to drink despite harmful consequences.

Harmful drinking can be defined as a pattern of alcohol consumption that causes health problems, including psychological problems such as depression, alcohol-related accidents, or physical illness. The guidelines say that harmful drinkers can become alcohol dependent, characterized by craving, tolerance, a preoccupation with alcohol, and continued drinking in spite of harmful consequences.

In my definition of an alcoholic, I'd include someone who either gets into trouble when they're drunk, or who has withdrawal symptoms such as morning shakes that require alcohol to stop, or who binges when they don't want to, or who drinks secretly.

Outside the medical community, people use the term "functioning alcoholic." But I'd say this simply means someone whose alcohol use hasn't led to them losing their job. There's a myth that all alcoholics are lying in the gutter. But a vast number are getting by, just not functioning as well as they could. And a proportion of those people are on the slippery slope to not functioning at all.

In the past, there has been a lot of controversy over the terms alcoholic and addict. Historically, those were the words used for people addicted to alcohol. Then from around the 1960s onward, it began to be thought that these terms were stigmatizing, that they were self-defeating, that they put people into a box from which they couldn't escape.

The new terminology was "dependence," and the condition was "drug or alcohol dependence syndrome." The basic idea was, if you got a physical withdrawal reaction from a substance, you were dependent. However, a lot of substances, including medicines, cause a rebound reaction when you stop them, and so technically cause dependence. Under such a broad definition, you could argue, for example, that you are dependent on antidepressants. But is that really dependence in quite the same way as alcohol?

I would argue not. I don't think this is a helpful way to describe addiction to alcohol. It doesn't take account of the psychological side, including the craving and desire.

Now we talk about both physical and psychological dependence. It's known that some substances—including alcohol—can be addictive for some people without causing physical dependence. For example, some extreme bingers will absolutely have the desire to get drunk twice a week and won't be able to stop themselves because they are addicted to the high. But they won't go into withdrawal.

It's also known that with other drugs, such as benzodiazepines, you experience physical dependence because you go into withdrawal without them—but the difference is, you don't crave them.

My opinion now is that the idea of addiction needs to encompass both the psychological and physical. And that, when it comes to alcohol, people will have differing amounts of each of these.

So my definition of addiction is a behavioral disorder underpinned by changes in the brain, which leads to continued use of a drug or substance in the face of problems such as withdrawal, and that your use of that substance interferes with your family and social life and causes you personal harms. And to go back to my original point, many people who are addicted to alcohol call themselves addicts and/or alcoholics. If this term is good enough for them, it's good enough for me.

HAVE I GOT AN ALCOHOL PROBLEM?

Below are the kind of questions a psychiatrist would ask you to assess if you have a problem with drinking. If you answer yes to any of them, it doesn't mean you are addicted to or dependent on alcohol, but it does mean you may be moving toward that. If you think this might be you, your doctor is a good person to talk to about it.

- ☐ Do you drink more than you intend to on a regular basis?
- ☐ Have people criticized you for your drinking?
- ☐ Has people criticizing you for your drinking annoyed you?
- ☐ Have you got into trouble or missed work because of your drinking?
- ☐ Have you been ill because of your drinking?
- ☐ Have you ever had to drink in the morning to deal with the shakes or nausea?
- ☐ Do you think you're spending too much time drinking and doing less of the things you used to do?
- ☐ Are you spending more money on alcohol than you really should?

HOW DOES ADDICTION HAPPEN?

There are multiple stages toward addiction. And you don't need to be physically dependent—number four in this list—to be addicted.

1. People start drinking because they like the pleasure it gives them. They also like the fact it makes them feel different, gives them a buzz. A lot of people like to drink because it stops them from being nervous or anxious too, a typical effect of this drug.

2. Tolerance to the effects of alcohol rises. That means your body becomes adapted to the effects, so it's prepared for it and you can drink more without being impaired—but it also means you need to drink more to get the buzz. There's no rule about how fast tolerance happens—we all vary—but it's likely to happen over a few days of heavy drinking.

3. Over time, people get more preoccupied with drinking. They might make sure more of their social life is in the bar, go out every Friday night with friends to drink, start to look forward to drinking. They seek out events where alcohol is easily available and make it a bigger part of their life.

4. As they get more tolerant to alcohol, people may start to get physically dependent (although this doesn't happen to everyone). The sign of this is that they go into withdrawal after drinking. This means they wake up early in the morning feeling shaky, anxious, nauseous, depressed. This might sound like a hangover, and it's true, the symptoms are the same.

I know my tolerance develops pretty fast. When I was captain of London University's badminton team in 1974, we did a tour in Sweden. A back injury meant I couldn't compete, so I spent my time playing table tennis and drinking the vodka the Swedes had asked us to bring them from Berlin because it was so expensive in their country. One morning, after two days of drinking, I remember putting sugar on my cereal and realizing my hands were shaking. The alcohol had changed my brain; I'd developed tolerance and, as I hadn't drunk for a few hours, I was in withdrawal. So I stopped drinking for the rest of the tour. But if I had drunk that morning to stop the shakes, and then done it again the next day and the next day, I might have become dependent on alcohol.

It's when people start to drink to get rid of the symptoms of withdrawal that the downward spiral into alcohol dependence begins. If you need alcohol to keep going after you've drunk alcohol, you are physically addicted. Withdrawal is a physical phenomenon that proves you are dependent on alcohol.

Whether you get withdrawal or not, drinking can become very compulsive. In your brain, pathways of behavior are laid down by the release of dopamine—this is true of alcohol and cocaine. The important transition is when people go from choosing to use alcohol to having a compulsion to do it.

A lot of the newer research relates to the fact that once people are addicted, breaking out of that addiction can be extremely difficult because of these changes in the brain. In my team's work at Imperial College, we have looked at brain scans of alcoholics who are abstinent and found they have altered functions in the dopamine regions of the brain. We have also shown that some other neurotransmitters, especially GABA and endorphins, are changed too.

On scans, we can see the original motivation to use a drug starts in a part of the brain that has a degree of knowledge and choice. This isn't your conscious mind; it's called the drive or motive circuit, the part that recognizes alcohol's positive effects and then decides you're going to take it again.

Eventually, as the person continues to drink over weeks and months, this behavior is downloaded into the habit part of the brain, so it becomes hardwired. At that point, the behavior happens subconsciously, without you making a decision to do it. Now, it's very hard for your conscious, thinking brain to get in the way.

I had one patient in the US, a man age 27, who dried out in the clinic, then left and said he was going to stay abstinent by going to AA. A few weeks later, he was back in the clinic. I asked him, what happened? He said, "I don't know. I found myself in a booze shop, drinking a quart of bourbon. I don't remember paying for it. I didn't want to drink it. I have no idea how I got there."

That is an extreme example of the subconscious drive overwhelming the conscious mind. You may think it sounds unlikely, but I believe him and the other alcohol addicts who've had similar stories.

We can describe this process in terms of four brain circuits that underpin addiction:

1) The reward circuit: that is, the hit of dopamine or endorphins you get when you drink.
2) The memory circuit: how you remember that alcohol is rewarding.
3) The drive or motive circuit: the area that makes you want to eat, drink, and have sex.
4) The conscious mind, the highest part of your brain: this is the part that should have top-down control. But it can be overridden: when your friends say, "Come to the bar! It'll be fun," your conscious mind might think, *no, remember my terrible hangover last week?* But your drive circuit remembers the great time you had and makes you forget your hangover and agree to go.

As you get addicted, the memory of the pleasure that's been registered by the reward circuit gets stronger. This then turns up the volume on the drive circuit, which in turn is more able to override the conscious mind.

Normally, most of us don't do things we shouldn't because our conscious mind says no. But over time, chronic drinking damages the brain, so it becomes less stable, especially during withdrawal, and so leaves you with even less self-control.

TO TREAT OR PUNISH?

A hundred years ago, we used to lock people up to get them to stop drinking. The reason we don't now is because it's been shown that stopping drinking is easy; it's keeping sober that's so difficult.

There's an interesting debate currently about whether people should be forced into treatment. The argument goes: alcoholism is a mental illness, so why shouldn't we prioritize its treatment and treat people compulsorily, as we do people who have a psychotic disorder, for example, schizophrenia? The justification for actively treating that group is to stop them from potentially being dangerous to themselves or society. Chronic alcoholism is mostly dangerous for the alcoholic themselves—they lose 20 years off their life span—although it can also have a huge impact on their families and on society at large in terms of personal violence plus crime, cost to the health system, and so on.

There is one big difference: regular medication for schizophrenia works, but it is harder to keep people from drinking, as treatments for alcoholism aren't as effective. And forcing people to be abstinent doesn't work: even in prison, for example, inmates will go to the effort of making a mush of chewed-up bread and fruit or juice, fermenting it, then drinking it.

The other reason treatment for alcoholism isn't prioritized is that society, and particularly politicians, prefer to view alcoholism as a failure of will rather than a failure of the brain. This kind of conflation of morality and science happens in addiction more than any other branch of medicine. It is a way of de-medicalizing addiction.

The problem with this approach is that, though it saves money on addiction services, this results in a much greater cost burden to the people who have to pick up the pieces, for example, the police, prison services, and health-care providers.

SELF-HELP FOR OVERDRINKING

1) Be aware of your addiction and be honest with yourself about it. What most therapy aims to do is strengthen the prefrontal cortex, the part of your conscious mind that makes decisions. This is the first step.

2) Be aware of your triggers. Don't go to bars if you don't want to drink. Remove yourself from places you associate with alcohol. Don't socialize with the people you always drink with—unless they are happy to go out to a place where you can both avoid alcohol.

3) Work out what drives you to drink. If you can identify the motives you have for drinking, you can then target your intervention and replace what you are missing or need

in your life. Make a list of all the other things you can do. If you feel unable to control yourself, can you think of ways that you can rearrange your life so alcohol isn't so present? If you are stressed, it's worth trying out various ways to relax: meditation, mindfulness, self-hypnosis, relaxation and breathing techniques, yoga, or just going for a walk. It doesn't matter what it is, as long as it works for you.

4) If you have low mood or anxiety, would you benefit from an antidepressant or talking therapy? Proven natural mood lifters are exercise and spending time outside and being in nature. If you crave excitement or a buzz, what nondrinking activities might give you that buzz (that aren't also self-destructive, such as sex with strangers or gambling)?

5) Continually remember the benefits of not drinking. This means that, when you are confronted with the desire to drink, you won't have to work to remember the drawbacks. Make a list of all the positive benefits of not overindulging, as many as you can, and keep it with you.

6) Know the times you're going to be vulnerable and most likely to relapse. Research suggests there are three of these. The first is obvious: times of stress. Typical examples of this might be divorce, a death, losing your job, or work stress. If you are drinking every night to deal with the stress of work, you need to tackle the stress and/or alter the way you think about work. The second time is less obvious: times of success or celebration.

People often relapse when something good happens, for example, a promotion. The excitement leads to them believing they have "cracked it" or feeling that they deserve a drink. If you think about it, success is, in effect, a positive kind of stress because now you have to maintain this new status. The third is alterations in mood. For example, if you feel depressed or just unusually low after drinking, be very careful. Because it's easy to get into a vicious cycle of drinking to overcome depression. That goes for any kind of drinking you do to change your mood.

AM I BECOMING DEPENDENT?

There are some warning signs that you may be tipping over into dependence, meaning that the amount you are drinking has changed and is now changing your body and your brain. Treat alcohol seriously: monitor the amount you are drinking, and where and when you are doing it.

The issue is that becoming dependent can lead to other problems, whether they involve relationships, work, finances, accidents, being arrested, or drunk driving, to name a few. If any of the following sound like you, now is the time to take stock, stop, or cut back.

- You notice you can drink more than you used to. I'd say that's especially true if you've doubled your consumption in the past few months. Think about it like this: if you doubled your consumption of food, you'd be concerned.
- You realize you have started drinking by yourself. And especially if you are drinking alone to quell anxiety.

- Every night, you really look forward to drinking when you get home from work.
- You have begun to end a day's work by drinking. It's a habit rather than planned. For example, you find yourself opening a bottle or going to the bar without having thought about whether you really want a drink.
- You insist on finishing the bottle of wine. Bottles will keep, at least overnight. But you may find yourself rationalizing that the wine won't be as good the next day, or it will go off. At this point, you're beginning to drink for the alcohol rather than enjoying the taste.
- You have started drinking more than your partner. For example, you might be having three-quarters of the bottle of wine, and your partner is only drinking one quarter. Or you drink half a bottle, then open a second one just for you.
- Your partner or a friend or a relative points out that you become unpleasant at some point during drinking. (NB: even if you can easily drink a bottle of wine a night without it impacting your relationships, you're still at risk of health harms and/or dependence.)
- Your partner encourages you to drink. Or you and your partner collude in drinking—that is, you both encourage each other to open the second bottle. It's like the play *Who's Afraid of Virginia Woolf?*, where the relationship dysfunction of the main couple is reflected in the amount they drink together. If what you do together is drink, together you'll have no resistance either. You can often see this in couples who run bars or pubs together.
- You're the one at the bar, getting in the shots. You never need shots.

I remember my shock when I first went to a Scottish pub and they were chasing shots with beers. The more you mix your drinks, the harder it is to control the speed and amount you drink.

- You are often the first person to the bar—because you need the alcohol. It also reveals that you're the person who is prepared to make the most financial sacrifice for it.
- When the bar closes, you're the one who often says, "Let's go to a club!" Or you tend to extend your evening of drinking by moving from place to place.
- Drinking has become your hobby or the only way you socialize. This often happens with retirees or expats.
- You have regular blackouts. A blackout means you've drunk so much, you've switched off your conscious brain. Regular blackouts mean you've done this repeatedly and haven't learned from it.
- You regularly vomit after drinking. Vomiting is protective, as it keeps you from dying from alcohol poisoning. Whether you vomit relates to how fast you drink and how much alcohol is in your stomach. Nota bene: if you vomit in your sleep, it can be lethal.
- You drink in secret. It could be taking a hip flask or half bottles to work, maybe tucking them in a drawer. Or it could be solitary drinking at home when other people can't see you doing it. Do you hide your bottles or empties?
- You start drinking every day at lunchtime. This has become less socially acceptable at work, but you may be in your job partly because in that company culture, it's still usual or even required. It's very common to start drinking

early on vacation and at weddings and celebrations. But do you carry on with this behavior on normal days?

IS THERE SUCH A THING AS AN ADDICTIVE PERSONALITY?

There is no such thing as an addictive personality. But there are individuals who are more vulnerable to becoming addicted. Your sex will make a difference, as men are more likely to be dependent on alcohol. The latest US data found that about 30 million Americans made the diagnosis of alcohol use disorder (the new term for dependence).[3]

Whether you get addicted can depend on your brain circuitry—maybe it didn't get formed correctly or has been damaged, either by the drugs themselves or other injuries. People who are impulsive or anxious or have another mental health condition are more likely to become addicted. Addiction also often depends on your life circumstances; as you can imagine, the most vulnerable times are those of high stress: divorce, bereavement, and so on.

IS ADDICTION IN YOUR GENES?

One of the reasons I got interested in alcoholism was reading about Danish adoption studies, from the 1950s and '60s, on the children of alcoholics. It showed that these children's risk of becoming alcoholics as adults was the same whether they were brought up in a drinking family, a normal family, or an abstinent family. This strongly suggests some inherited basis for alcoholism rather than it being a behavior learned from the alcoholic parent.[4]

Since then, it's been shown that alcoholism does run in

families. If your parent has had or has alcoholism, your own risk is three to four times higher.[5]

But you don't just inherit alcoholism; in many cases, you inherit one or more of the various different traits that make it more likely you'll be addicted, for example, impulsivity or conduct disorder, as shown by a recent study on twins.[6] Some traits are mediated through a predisposition to other factors we've mentioned, for example, anxiety or stress.

Others relate to the effects of alcohol on the brain. And some relate to the way your body deals with alcohol when you drink it.

One of the most important findings on vulnerability to alcoholism came from research on the sons of alcoholics by Professor Marc Schuckit at the University of California at San Diego. It found male children of male alcoholics had significantly higher rates of alcoholism than the general population. And more detailed analysis of those boys found that they seemed to have inherited a tolerance to alcohol, possibly even from the first time they drank it.[7] This meant they were able to tolerate higher amounts than their peers. Yet despite being to some extent pre-tolerant, they were more likely to be alcoholic. It could be that they got some kind of prestige from being able to drink more, or that they just found it easy to drink more than their friends, and this made them vulnerable. Or it could be their heavier drinking produced other changes in their brain that then led to dependence and alcoholism.[8] I believe the trait being passed on, subsensitivity to alcohol, may be due to an alteration in the GABA system. The same kind of genetic factor appears to apply to women with a family history of alcoholism too, although it seems this effect is not as strong.[9]

You can also inherit a tendency to be less likely to become an alcoholic.

Probably the best known tendency for protection comes from the alcohol flushing reaction. One estimate is that 70 percent of East Asians (Japanese, Chinese, Taiwanese, and Korean ethnic groups) have this, and 5 percent of European ethnic groups.[10]

And we know that Chinese people have lower rates of alcoholism and reduced rates of the toxic effects of alcohol, such as cirrhosis and liver cancer too.

If you have this trait, you will have a variant of the gene aldehyde dehydrogenase (ALDH) that slows down the breakdown of the alcohol by-product acetaldehyde, which is a poison. This means that if you drink, you get high blood levels of acetaldehyde—which is what makes you flush and sweat, as well as gives you a headache, nausea, and speeds up your heart rate. The alcohol treatment disulfiram (Antabuse) also blocks this enzyme, as described in the Treatment section in this chapter. If you take it and drink, you get the same unpleasant effects—the flushing reaction—which (the point is) can help you resist the temptation to have another drink.

You can get the same effects from taking certain drugs, including the antibiotic metronidazole (Flagyl), which is why it has a warning on it not to drink. And for anyone who's had a drink while taking it, I'm sure the experience has taught them one thing: to listen to their doctor!

I experienced this effect myself many years ago when, before a formal dinner for neurologists at St. Mary's Hospital, I took quite a big dose of phenylbutazone (Bute) for my arthritis pain. At the time, it was quite a popular antirheumatic (it's still used for horses). I couldn't understand why I was so inappropriately and illogically drunk, flushed, and loud from a normal level of formal dinner drinking. My girlfriend picked me up and said I was more outrageous than normal. I was even sick afterward. I looked it up the next day and discovered that phenylbutazone also blocks ALDH. It wasn't all my fault, which is probably why she agreed to forget the episode and marry me. . . . Wikipedia has a good entry on the many other drugs that can block ALDH—if you're prescribed any medication, it's worth reading before you drink![11]

Each gene has two parts, or alleles. If they are both defective for ALDH, you will get the flushing reaction very badly and will almost certainly learn to avoid drinking. If only one is defective, you will have just some of the enzyme and so may be able to force yourself to put up with the reaction, as it won't be so extreme. The downside is that if you do drink, you will have more toxic acetaldehyde going around your body for longer, and research shows that will put you at a higher risk for cancer, in particular, esophageal cancer.[12]

Recent advances in technology have made it possible to measure changes in all the genes in the body—this is called whole genome analysis. Analyzing data from about half a million people, a team from Imperial College London found 46 new genetic markers linked to alcohol intake, which together accounted for 7 percent of the variation in people's total drinking. The team also identified genetic pathways shared between alcohol intake and brain networks associated with psychiatric disorders such as schizophrenia.[13]

WHAT CAN LEAD TO ADDICTION

Anxiety can lead to addiction: Imagine you are a young man who gets nervous about, for example, talking to girls. When you go out, you drink because you find it reduces your anxiety. This happens because alcohol increases your levels of the neurotransmitter GABA and decreases your levels of gluta-mate, so you feel relaxed.

If you drink regularly, that is, most days, then your toler-ance will start to build, and there will be a physical change in your brain. In response to the increase in levels of GABA, your GABA system becomes down-regulated. At the same time, your glutamate system is up-regulated. Now, you have to drink more to get the same level of anxiety reduction.

Also, you'll find that when you stop drinking, you feel even more anxious. You now have a hyperexcitable brain that needs more alcohol for you to feel normal. You reach a point where if you don't drink you are too anxious to function. You are now a dependent drinker.

Having fun can also lead to addiction; some people get a huge burst of pleasure from drinking—an endorphin rush. As your brain gets tolerant to the increased levels of GABA, it's also becoming sensitized to the increased dopamine and endorphins.

In this case, the more you drink, the bigger the buzz you get. One of the reasons people who respond like this lose control over their drinking is that as the buzz increases, they drink more. Their dopamine system gets turned on, so they continue to seek and desire alcohol and enjoy it—even as they become more tolerant to its effects. They are motivated toward drinking even if they don't want to.

Another reason people turn to alcohol is, as well as reduce physical pain, it gets rid of emotional pain—it takes away irritation, frustration, and tension. In fact, it acts on the brain's emotional circuits in a way similar to opiates. Some alcoholics report that they feel numb and protected when they are drunk, that life feels like a better place. There are all sorts of reasons for this kind of pain that are beyond the scope of this book, from illness and poverty to disability and loss. It's important to remember that, for a lot of people, life is tough, and the world is hard and challenging. For some, alcohol relieves this pain.

THE FAST TEST

This is one of the initial tests a health practitioner can use to assess your drinking.

	0	1	2	3	4	Your Score
How often have you had 4 or more drinks if female, or 6 or more if male, on a single occasion in the last year?	Never	Less than monthly	Monthly	Weekly	Daily or almost daily	

ADDICTION: HAVE I GOT AN ALCOHOL PROBLEM?

	0	1	2	3	4	Your Score
How often during the last year have you failed to do what was normally expected from you because of your drinking?	Never	Less than monthly	Monthly	Weekly	Daily or almost daily	

	0	1	2	3	4	Your Score
How often during the last year have you been unable to remember what happened the night before because you had been drinking?	Never	Less than monthly	Monthly	Weekly	Daily or almost daily	

	0	1	2	3	4	Your Score
Has a relative or friend, doctor, or other health worker been concerned about your drinking or suggested that you cut down?	No		Yes, but not in the last year		Yes, during the last year	

An overall total score of 3 or more on the first or all 4 questions is FAST positive.

Alternatively, here is an even shorter questionnaire a health practitioner can also use:

Do you regularly have more than six drinks in one sitting? Or do you regret a drunken escapade that took place in the past year?

If the answer is yes to either, take the full FAST questionnaire.

SO YOU THINK YOU HAVE A PROBLEM . . .

You may know you have a problem with alcohol. Or you may suspect it, after reading this book. Or it may be your problem has been recognized in the doctor's office. As the health harms of alcohol are now so well recognized, all doctors are encouraged to ask questions about drinking.

The FAST questionnaire is a selection of questions from a longer test doctors can use, called AUDIT.

If the result of your FAST/AUDIT assessment is that you drink too much, your doctor will give you advice on cutting down to sensible levels or on giving up drinking altogether. Getting your doctor's advice may be just what you need to do this; it is for many people—pioneering research by Professor Griffith Edwards, a UK psychiatrist and a leader of treatment research for alcohol dependence in the 1960s and 1970s, showed that firm words from a doctor can encourage people to cut down.[14] Since then there have been many studies of what has become known as a "brief intervention" in primary care, an evidence-based, structured conversation, showing these to have real benefit. These are designed not to tell you off, but to help you see the risks of your drinking and the benefits of reducing and to motivate you to change.[15]

Your doctor may also ask further questions to find out if you are dependent on alcohol.

If you try to cut down and find you can't, because of either physical or psychological symptoms, or if your doctor assesses that you are dependent, the next step is to be referred to specialist alcohol services.

DETOX: WHAT HAPPENS?

When you stop drinking, you go through a detox, which simply means physical withdrawal from alcohol.

If you have tried to do this on your own but found it too hard because the cravings were too strong or you felt very ill or anxious, you will be offered medical assistance. There is a very well-established reduction protocol for detoxing, usually done at home with medication prescribed by your doctor. Much less often, it is done in a hospital or at a private rehab center.

With the help of medication, you should be able to come off alcohol safely and without a great deal of distress. Medication will stop unpleasant symptoms caused by the deficiency of your GABA system, such as anxiety and visual hallucinations and, more important, life-threatening ones such as seizures or delirium tremens during withdrawal.

During a detox, you are transferred from alcohol to another drug, usually the benzodiazepine Valium or Librium. Then, over the course of a week, the medication is reduced slowly, allowing your brain to adapt and recover. You do lose a week and it's not exactly enjoyable, but the main thing is, it prevents any serious medical events. There is a downside: some people do find it hard to come off the anti-withdrawal medication.

TREATMENT

After the detox, you'll be offered specialist treatment with addiction services. You may also be offered this by your doctor if you are not dependent but you have had a brief intervention and it hasn't worked. In most cases, if you are considered needy enough, treatment will involve either one-to-one or group psychological therapies.

MEDICATION

There are medications to help people reduce their drinking or stop, but very few people are offered them. Even though doctors see so many alcohol-related problems, sadly very few have been properly educated about what and how to prescribe.

The same is true in specialist addiction services, despite the fact that doctors there do prescribe drugs to treat addictions, for example, for heroin and tobacco. It's worth raising the possibility of medication with your therapist—these are licensed medicines with proven efficacy in controlled clinical trials. They are used much more widely in the private sector.

Drugs are also much more widely prescribed in other countries. In the US, you might be prescribed naltrexone, whereas if you lived in Italy or Austria, you would likely be prescribed sodium oxybate (see page 164). If you lived in France, there's a very good chance you'd be put on baclofen. In Britain you may be prescribed nalmefene, which is used in people who are still drinking to help reduce bingeing, or acamprosate, which helps prevent relapse in people who've stopped.

1. **ACAMPROSATE (CAMPRAL)** This has been around for nearly twenty years. It's an anti-glutamate drug, which means it reduces cravings for alcohol and possibly also symptoms of withdrawal. It's been shown to work—although not in everyone—and to be cost effective, but it's underused for reasons I don't understand. Sometimes it is used in combination with Antabuse.

2. **NALMEFENE (SELINCRO)** This is approved in Europe and other regions, but not in the US, to help reduce alcohol consumption in people who binge drink. It works on reducing the pleasure of alcohol, like naltrexone (see this page). Because most GPs in the UK aren't educated in alcohol-related problems, it is only very rarely prescribed, despite commissioning groups being told they should support it. It's widely used in France.

3. **DISULFIRAM (ANTABUSE)** This makes people vomit and flush and feel very ill with chest pain and a fast heartbeat when they drink, by blocking an enzyme needed for the body to process alcohol. People used to be prescribed Antabuse as part of aversion therapy. The idea was, after they had relapsed and experienced the results, the next time they were tempted to drink they would remember how ill they'd been. The problem is that your memories of unpleasant experiences tend to fade, and the desire for alcohol can overwhelm them. Consumption also has to be supervised as, if you don't take it, you won't get the (desired) bad effects.

4. **NALTREXONE** Designed to reduce relapse in abstinence, like nalmefene, it works by blocking the endorphin system. This takes away the pleasure of drinking and so can help

people who lapse avoid a full relapse. It is also used to treat opioid misuse.

5. **SODIUM OXYBATE (ALCOVER)** This is a very interesting drug, discovered in Italy by a great pharmacologist, Professor Gian Luigi Gessa. He bred a strain of rat called the Sardinian alcohol-preferring rat, then looked for drugs that would stop them from preferring alcohol. He first showed that sodium oxybate stopped rats from drinking as well as prevented them from having withdrawal symptoms, then demonstrated that the same happened in humans.[16]

Not yet licensed in the US or the UK, sodium oxybate can be used in private clinics on a named-patient basis. It's been used in Italy for 20 years and in Austria for nearly that long and it's currently (as of writing in 2020) under review by the European Medicines Agency. There's been a movement for several years to get it licensed in Britain, but it has failed. One reason is that, in its original, liquid form, sodium oxybate is used recreationally; it has the street name of GHB. So while this drug would help alcoholics, it's currently denied them because of the fear that it might be abused.

The other argument against it is that its effects are too much like those of alcohol, so someone taking it is swapping alcohol for another dependency syndrome. The best analogy here is between vaping and cigarettes. As a pragmatic harm reductionist, I say being dependent on the nicotine in vaping isn't going to harm you nearly as much as smoking cigarettes. In a similar way, being dependent on sodium oxybate is going to harm you way less than alcohol.

Sodium oxybate has a newer formulation (Alcover), an orally dissolvable powder, which no longer gives users a buzz. There is good evidence that Alcover helps very heavy drinkers, people who are in the last decade of their life because of alcohol. As the only drug strong enough to compete with alcohol, it could be lifesaving for so many.

6. BACLOFEN Originally developed in the 1960s to treat spasticity, this drug was discovered as a treatment for alcohol by a famous French-American cardiologist Dr. Olivier Ameisen, an alcoholic who wanted to treat his own illness. When he started to take baclofen, which works on the GABA system, he found he was able to stop drinking.

He wrote a book, *The End of My Addiction*, about how he cured his alcoholism. It became a bestseller in France, but there isn't as yet any good trial evidence that baclofen works. This might be because you can get tolerant to it, so people are often given too low a dose. It can be prescribed in the US but only off-label.

PSYCHOLOGICAL TREATMENTS FOR ADDICTION
Alcoholics Anonymous (AA)

The most famous talking treatment of all, AA is effectively group therapy plus multiple other elements. The 12-step program is a sophisticated psychological process made simple. It covers your life history of drinking, so it gets you to work through how and why you drank as well as the problems you caused for yourself and others. It helps you get insight and understanding into your reasons for drinking. It then helps you develop mental strategies that support your abstinence.

The group element adds peer pressure, and you also have a buddy—your sponsor—on call who has been through the whole process. Plus there's lifetime follow-up. But the greatest benefit is that you can find a meeting in every major town in the US.

Many people in AA stay sober for the rest of their lives. The drinkers who I've seen that don't find it helpful are those drinking to deal with an anxiety disorder because they find it intimidating to be in a group.

Motivational interviewing

The aim of this therapy is for people to understand the reasons for their drinking, then develop strategies to overcome them. It's essentially about replacing the pleasures and rewards of and the desires for alcohol. It helps you reframe your thinking so you can get similar or more pleasure from not drinking.

For example, if you drink to deal with work stress, you will learn new ways to reduce stress. And you'll develop strategies to help you make the tough decision not to drink after work. It will help you find new evening activities, such as a sport, so you don't find yourself in the same old drinking haunts. It will also help you think through what you can do at a work function when you'd usually drink. Or, if you often drink and then argue with your partner, it will teach you to reframe this as: if I don't drink, I will have a better relationship.

Cognitive Behavioral Therapy (CBT)

The cognitive part is to help you understand the thinking processes that are behind drinking. As people often engage in behaviors without thinking, analyzing them helps you gain mastery not only over them but also over their associated thoughts.

There are four elements to CBT. First, you work out the factors that lead to you drinking. You look at what is happening internally: Is it to deal with anxiety or depression, or is it because you crave alcohol? Then you look at the cognitive processes related to that: What are your expectations of drinking? Do you have a good time? Are you denying the harms? You also focus on the consequences of your drinking. For example, do you remember the fun parts and forget the fact that you threw up or were arrested?

WHAT TREATMENT SHOULD I TRY?

The only thing I'm confident of clinically is that people who have any underlying psychological or psychiatric problem, such as depression, anxiety, PTSD, or bipolar disorder, who then become dependent on alcohol, should have these problems treated first (see Chapter 4 on mental health).

After that, treatment is a question of trial and error. We don't yet know how to identify the people who will do best on any one drug or talking treatment; all we can do is give them what we have and see if it works.

A US study in the 1990s called Project MATCH, which cost $27 million, matched people with different talking therapies. The idea was to find out if that would increase the success rate.

One group had Cognitive-Behavioral Coping Skills Therapy, which helps people to deal with poor self-esteem and thoughts of failure. The second group had Motivational Enhancement Therapy, which helps people to build on their strengths. The third group did a 12-step program derived from AA. After 12 weeks, the results showed that you could not predict which program would work; in fact, they all worked equally well.[17]

FLASHPOINTS FOR OVERDRINKING

As you've read, it's easy to tip over from being a normal drinker into one who is dependent. These are the times of your life when you are most vulnerable.

- **Divorce or a breakup**

  When people split from a partner, it's a cliché, but they can end up engaging in solitary drinking to numb the emotional pain. You might have heard the urban myth about the woman who was found dead in her flat by the police and how they also found hundreds of wine bottles and hundreds of uneaten pizzas. The story goes that she'd sit at home and order a bottle of wine and a pizza to be delivered . . . but never eat the pizza.

  Because most of our socializing is done around alcohol, if the newly single want to meet people, they may feel under pressure to drink. And then there's building your liquid courage for dating too. But the main effect is thought to be that, when you have a partner, they look out for your health, including if you are drinking too much. A Swedish study showed that divorce leads to a significantly higher risk of being diagnosed with alcohol-related disorders, but that remarriage lowers that risk.[18]

• Going to college

In the US, despite the drinking age being 21 years, drinking at college has become a ritual that students often see as an integral part of their higher education experience according to the NIH. Many students come to college with established drinking habits, and the college environment can exacerbate the problem. According to a national survey, almost 55 percent of college students ages 18–22 drank alcohol in the past month, and more than 1 out of 3 of them engaged in binge drinking during that same time frame. About 1,825 college students between the ages of 18 and 24 die from alcohol-related unintentional injuries, including motor-vehicle crashes.[19] Tragically, this does kill young people every year: take the case of Ed Farmer, a first-year UK economics student who died after an initiation ceremony with the university's Agricultural Society, which involved a bar crawl and 100 triple vodka shots.[20]

• Death of a child or a loved one

People use alcohol as part of their coping strategies, to deaden the pain of loss. It seems this association is highest two years after bereavement, according to a Hungarian study. At this time, the proportion of men whose levels of drinking put them at risk of health harms reached nearly 30 percent.[21]

• Job

It's not unusual to drink to deal with work stress. But the biggest risk comes when people lose their job and find themselves with days and days of free time and less reason to get up in the morning.

- **Retirement**

  The number of older people dependent on alcohol or who drink heavily has risen fast. It's a consequence of an older population but also what happens when people retire and find themselves with disposable income and with more time to drink and no reason to put on the brakes. Added to that, due to other illnesses and changes in metabolism and medications, it's likely that older adults are at greater risk from alcohol-related harms.

- **Moving abroad**

  Similarly to retirement, people often end up with a lot of free time on their hands. The social life of the expat may be based around drinking, and booze is often cheaper too.

WHAT CAN YOU DO IF SOMEONE YOU LOVE DRINKS TOO MUCH?

If you want to talk to someone about their drinking, don't do it when they are drunk. It's completely pointless; people can't properly engage their brain when they are intoxicated. And chances are they won't react well.

We know that pressure plus support from family and loved ones can be a powerful way to get people to want to come off alcohol. However, a doctor can't make a spouse or partner go along to a meeting; the alcoholic has to make that decision. If you are the partner or relative of someone who drinks too much and can get your loved one to talk to you about alcohol, it's a good start.

WHAT TO DO IF YOUR PARTNER IS DRINKING TOO MUCH

A lot of people would refuse to start a relationship with someone who smokes, and I'd say it'd be sensible to do the same for someone who drinks too much. Obviously this isn't practical advice for a relationship at a later stage.

If your partner has mentioned that they are worried about drinking too much, it's a good idea to monitor their drinking, the same way you might observe and comment on what they are eating if they had told you they wanted to cut back. That way, you can be the nudge for them to slow down or stop. By doing this, you can hopefully help them before they become alcoholic.

However, I'm assuming that if you're reading this, you are already past this point.

If you are worried, the first thing to do would be to note what they are drinking over a few weeks, and any adverse consequences. And at some point—when they are not drunk, you have the time, and you are alone with them—have a conversation about it.

If your partner says they understand, together you can develop a strategy to reduce the harms of their drinking (see Chapter 9 for details of how you might do this). It might be that you agree to set nights to drink or a set limit of drinks per night.

If they need convincing, it's worth giving them evidence of their altered behavior; for example, you could film them. You can talk through the problems that their drinking is causing, whether it is destroying your relationship, or their work, or their relationship with any children.

Rather than get angry, try to be calm and clear in your analysis of their problem. There's a real opportunity, during the contrite phase after drunkenness, to get them to see that controlling their drinking is not only in your interest but in theirs too. Perhaps you could ask them to read this book, for more reasons why it's a good idea to stop or cut down their drinking?

You could also ask them to make an appointment with their doctor for a health checkup. Doctors are trained to ask about and pick up on harmful drinking, and a diagnosis of high blood pressure or an abnormal result of a liver test or scan can be a powerful motivator. If they have an underlying psychological issue that is leading them to drink—such as bipolar disorder, depression, anxiety, or OCD—encourage them to have treatment for that.

If the issue is stress at work, you can suggest alternative stress-reducing behaviors.

If it's more your partner's social behavior when drunk, counseling might be a good idea, either together or for them alone, so you can have an independent person helping you work through the issues.

You don't have to leave someone just because they are an alcoholic. After all, it is a mental illness, like depression, and most people wouldn't leave someone just because of that. Many people can and do overcome their addiction. But if your partner is violent or refuses to seek treatment and it is affecting your or your children's lives, it may be time to leave. As well as your own doctor, another place you might want to try is Al-Anon, a group with a 12-step program that exists to support people through the problems of being in a relationship with an alcoholic.

WHAT TO DO IF YOUR CHILD IS DRINKING TOO MUCH

This is challenging because you won't know how much your child is drinking if they are drinking socially, separately from you. But they are probably drinking too much if they are coming home drunk and vomiting, and/or having terrible hangovers. Especially if they are doing this regularly.

Often an issue for teenagers or young adults is that they'll say they are only doing what their friends do. It's your job to explain that doesn't make it safe. You need to base the discussion—assuming they will have one with you—on reason and evidence. You could use some of the facts in this book or ask them to read it themselves.

You could also start by asking them: Might you be drinking more than your friends? How do you feel after you've been out drinking?

Explore ways that life might be better if they didn't go out and get drunk. Could they do more of a sport they love? Or another activity? Could you do something together with them in your free time?

If your child refuses to engage in conversation, it is very hard. Can you get support? There are expert organizations for this (see the Resources section at the back of this book). At this point, if you are giving them money and they are spending it on alcohol, stop. Also let their doctor know, if they are below the age of consent.

WHAT TO DO IF YOUR PARENT IS DRINKING TOO MUCH

If you are a child in this situation, please speak to someone. Can you talk to a teacher? Or a relative? You could call the

National Association for Children of Alcoholics (NACOA) (888) 55-4COAS. You are not the only one: we know roughly one in ten adults are alcohol dependent, so that must mean that about one in ten parents are too. We know there are schoolchildren who have to look after their drunk parents. And we know that, for some children, that is what happens every day.

When I took part in a *You and Yours* phone call-in radio program on the BBC, the topic of how other people's drinking had affected the callers' lives got their largest ever number of callers at the time. Many were older people who were still distraught and distressed by what had happened to them at the hands of a parent many decades before.

If you are an adult with a parent who is drinking too much, see the advice given previously for partners. You can also seek help from the organizations mentioned on the previous page.

WHAT TO DO IF YOUR FRIEND OR RELATIVE IS DRINKING TOO MUCH

All the advice given in the section on partners applies here too. This can be particularly hard because you may not be living with that person, and they may be less willing to speak to you. In fact, you may find that they resist talking to you and deny that drinking is affecting them. They may also turn against you and end your friendship or contact; at least you'll have done the right thing. And if in the future they reform their behavior, they will likely return to thank you for helping initiate this change.

See the Resources section at the back of the book for organizations that can help.

COMMON QUESTIONS

SHOULD I WORRY IF I'M ALWAYS THE LAST ONE LEFT AT THE PARTY?

Some people don't seem to know when to stop. And it may be their brains are different from the norm. For them, the endorphin rush becomes so pleasurable that they lose the ability to make the decision to stop drinking, and just keep seeking the rush. Even though they may start off the evening wanting to drink normally, the alcohol changes how their brains work.

The genetics behind being more likely to binge drink isn't known. What we do know is that if you are a binge drinker who loses control, you can bring yourself back toward normality by blocking the endorphin system. The drug nalmefene may help, as it works to stop the effect of the endorphin release. Try controlling or stopping your drinking (see Chapter 9), and if this doesn't work, see your doctor.

I STILL GET UP IN THE MORNING AND GO TO WORK, SO HOW CAN I BE AN ALCOHOLIC?

A critical part of a diagnosis for any mental disorder is impairment. Alcohol affects your judgment; you can be impaired and not know it, so you might not be the best judge of whether you're an alcoholic. I'm afraid that going to work simply means you go to work.

Ask your partner, a trusted friend, or a family member to tell you the truth about your drinking. This will give you your best chance of being confronted with the reality of your behavior.

WHY WON'T MY PARTNER EVEN DISCUSS THEIR DRINKING PROBLEM, LET ALONE TALK ABOUT STOPPING?

It's common for people to be in denial about a drinking problem. They don't want to believe that alcohol is an issue because they need it. Most people think their judgment when under the influence of alcohol is OK, or even that it's better than when they're sober.

There are very few people who consciously want to be an alcoholic and destroy their lives. After all, everyone can stop drinking; you just don't pick up the bottle. The reality is that people who are alcohol dependent can't stop because their brain won't let them stop. When someone is at that point, it's not an intellectual decision. Their brain is beyond reason and has modified so that alcohol has become the driving force of their life. People who are physically dependent can even have drinking dreams, as well as think about alcohol all the time they're awake.

There is a tension that exists with drug use, that the pleasures outweigh the harms. The most common reason behind someone ending up in treatment is that they've lost something: their job, partner, kids, or their liberty. People often only make the decision that they want to change when their life reaches crisis point and they finally realize that the alcohol has contributed to or caused it.

Some people will still refuse treatment. One extreme example of this was a 50-year-old Roman Catholic priest, an intelligent man and a university graduate.

When I met him he was living in a Salvation Army hostel, having dropped out of the priesthood ten years previously due to his drinking. It was very sad to hear how this man's life had slowly been destroyed by alcohol.

The priest's drinking behavior was not untypical of an end-stage alcoholic. He got a weekly check every Thursday, for around £150 (approximately US $200). He'd cash it and go to the pub. He'd drink half of it on Thursday night, the other half on Friday. Then, for the whole of Saturday and Sunday, he'd be in withdrawal.

As we talked, we agreed this was a terrible way to live. I told him about the proven medications and treatments for alcohol. He said, "That's very kind of you, but I don't think so." His love of alcohol was greater than his love of life or even of God. The pleasure he got from two days of drinking a week outweighed every other pleasure in his life.

His story is extreme but what it illustrates is, first, the person has to want to change. And second, in order to overcome the addiction to alcohol, they've got to put a lot of good reasons not to drink into their brain, as well as develop the skills and strategies to overcome the subconscious drive to drink. This will take expert help and possibly medication. Your partner won't stop drinking because, at the moment, they can't.

THE SOCIAL BENEFITS
OF ALCOHOL

HUMAN BEINGS ARE social creatures. And you could say alcohol is THE social drug. Of all the recreational substances we take, perhaps with the exception of ecstasy, alcohol creates the greatest sense of sociability. You could say alcohol provides the perfect drug cocktail for socializing: you become energetic and positive as if you're on cocaine, you become relaxed as if you're on marijuana, and you love other people as if you're on ecstasy.

There has been a lot of work done to analyze the harms of alcohol, for obvious reasons, but there hasn't been nearly so much scientific scrutiny of its benefits. These are both social, such as relaxing, bonding, and spending time with other people, and creative, allowing us to expand our minds into new thoughts and territories.

These kinds of benefits are hard to measure. But I'd say they had and still have an enormous value in society. In fact, human beings have been making alcohol and drinking it together for thousands of years. It's the easiest drug to make— all you need is rotting fruit.

There's an anthropological theory that alcohol was a crucial element in the development of organized societies, in our species making the switch from nomadic hunter-gathering to farming. It says that crop cultivation may have begun in order to grow wheat to make alcohol as well as for food.[1]

It's also been suggested that alcohol, as a driving force behind human creativity, has fueled the development of cultural expression, from art to music to religion.

And it's true there's increasing evidence that every society in history had a way of making alcohol, and that it was used at significant events like rituals and for special occasions. It's only in the past millennium that it has been banned by some religious groups, for example, in Islam.

Honey, for example, was fermented into a strong, sweet drink called mead. Some experts think this may even have pre-dated the fermenting of wheat, that people were drinking mead in Africa over 40,000 years ago.[2] It's thought mead became popular in northern Europe because it was harder to grow wheat there. In the early Middle Ages, it reputedly fueled Viking bonding ceremonies and warfare (see Chapter 3), a bit like alcohol does for many sports teams today.

Distillation of wine into spirits came much later, probably in Arabia about AD 900. In fact, the word alcohol is derived from the Arabic *al-kuhl*, meaning "body-eating spirit" (for some people, this is sadly true). The distillation technology reached Europe, where in the thirteenth century Arnaldus de Villanova, a physician, commented: "We call it *aqua vitae* [water of life], and this name is remarkably suitable, since it is really a water of immortality. It prolongs life, clears away ill-humors, revives the heart, and maintains youth."[3] We know more now, of course.

But, for most of us, alcohol does have a broadly pleasurable effect.

A group of French researchers set out to measure the benefits and harms of alcohol compared to other addictive substances, using a similar method of analysis to those done on harms only by my group.[4]

They asked addiction experts to put a value on the benefits we get from the most commonly used addictive substances and our behavior (including alcohol, tobacco, cannabis, cocaine, heroin, amphetamines, ecstasy, and gambling) in order of their impact.

The researchers broke the benefits down into six categories.

1. Hedonistic benefits. The intensity of the pleasure obtained.
2. Identity benefits. How it allows an individual to socialize and be part of a group, plus its cultural value.
3. Auto-therapeutic benefits. How it soothes internal suffering and tension.
4. Economic benefits. The value of production, sales, consumption, and so on, to society.
5. Social benefits. How it helps maintain social balance.
6. Cultural benefits. How it's used for celebrations and social rituals.

As in my group's analysis, alcohol came out as the most harmful addictive substance to society. But it was also the one perceived by the French researchers to have the most benefits. In fact, it was so much better than the other drugs and activities, they called it an outlier.

My question to you is, what benefits does alcohol bring to your life? Nearly all of our celebration rituals revolve around alcohol, from the cradle to the grave. We wet the baby's head, we toast exam results and birthdays, we crack open the bubbly at both wedding breakfasts and divorce parties. And we say good-bye to our loved ones with alcohol at a funeral wake (and drown our sorrows too). When you are working out how much to drink—or not drink—you may want to factor in the social benefits you get from it.

DRINKING: THE GOOD TIMES, OR WHY WE DRINK

The most common reason people give for drinking is having a good time with others. How much of a good time was measured in a study that used an app called Mappiness. The app asks you to measure how happy you are on a scale of zero to 100, then to plug in where you are, what you're doing, and who you're doing it with. While they were drinking, people reported being an average of four points happier.

Another app-based survey by Pebble (a now discontinued smartwatch) confirms what we know about the feel-good glow of drinking. It asked people for their self-rated level of happiness as well as what they'd done in the previous hour, and drinking came out as giving the best buzz—followed by yoga, exercise, socializing, meditation, food, then caffeine. It scored pretty highly as being energizing too.[5]

People also use drink to change their mood, to reduce negative emotions, and provoke positive ones, at least in the short term. This isn't surprising, considering alcohol's endorphin and dopamine brain hit.

Alcohol lubricates our interactions with others too. If you feel nervous about talking to people at a party, as soon as you have your first drink, it helps you overcome that threshold of anxiety so you can talk to people. A glass to hold gives us something to do with our hands in awkward situations, particularly now that smoking has become so vilified. Go out for dinner and you add even more of the good things, like socializing and food, into the mix too.

There's another level to this. Research shows that people who drink socially—especially those who have a regular local watering hole—tend to have more friends and so more emotional support, a key source of mental health. They also feel more contented and more involved in their local community.[6]

Alcohol is a major part of how we bond. Asking someone to go to the bar is a culturally acceptable way to make friends, especially for men. Half the reason why people spend their evenings or weekends playing in sports teams is probably the bonding in the bar after the game or match.

Drinking has also been shown to promote social feedback: facial expressions of interest, for example, as well as verbal ones. So you look more animated, more interested, and you feel as if you're being more interesting (hopefully you're not the boring drunk, droning on!).

In a study, when strangers were put together in groups of three, those given alcohol were more likely to smile fully, and smile at the same time as one another. The drinking groups were also more likely to have everyone joining in the discussion.[7]

Drinking may make you less aware of other people's criticism. But it may also make you feel that people are more positively disposed toward you, at the same time as you become more positively disposed to others too. Good feelings all around![8,9]

Research suggests that drinking in groups—in particular binge drinking—takes the positive feelings up a level, to euphoria. A prime example of this are bachelor and bachelorette parties.

THERE IS A DOWNSIDE

While people like to drink together, it's easy to drink too much. One study found that there were four main motives for drinking more in a group: copying (automatically copying another's drinking behavior and how much they drink), conformity (fitting in with the group), hedonism (escape and having a good time), and winding down (relaxation). And once we've used alcohol as a stress valve, that behavior becomes embedded.[10]

IS YOUR BOOZE BUZZ IN YOUR HEAD?

Something that might make you think—and especially for anyone who wants to stop drinking—is that you can get the buzz without the booze. In a study in New Zealand, people were put in a room that looked like a bar. Half of them were told they were drinking vodka and tonic, half just tonic. But, in fact they'd all been given tonic. The fake drink glasses were rimmed with limes dunked in vodka so they smelt of alcohol, but the alcohol content was negligible. The—totally sober—people acted drunk by flirting and giggling and, even when told otherwise, believed they had been drinking.[11]

For many people, in many circumstances, alcohol does what you want or need it to do. If you want to be happy and have fun, it can do that. If you want to numb your feelings, it can do that. If you want to bond with the group and let loose, it can do that too.[12,13]

DOES ALCOHOL CHANGE YOUR PERSONALITY?

How does your behavior change when you drink? Possibly less than you think. In a lab study, groups of three or four people were given alcoholic or nonalcoholic drinks, then asked to complete puzzles or discuss questions together. The tasks were designed to help reveal their personality traits.

When people were asked to rate themselves, they thought that after alcohol they became less self-disciplined, less open to trying new things, less agreeable, but more emotionally stable and extroverted.

But when the rest of the group rated them, the only change the others noticed was a boost in extrovert traits—that they were more gregarious, active, and assertive. The conclusion? Alcohol doesn't change you as much as you think.[14]

If people do transform under the influence—and we've all seen people who cry or get angry when drunk—it may be because alcohol's disinhibiting effect is allowing any repressed emotions to come to the surface. Most of the time we try to control our behavior, to keep our emotions in check in a way that feels socially appropriate. When you drink, you can just let it all hang out.

So if you're miserable at work but putting on a brave face, if you drink you may end up upset, "drowning your sorrows" or at least sharing them with a friendly and receptive audience!

Or alcohol might dissolve your facade and liberate your underlying aggression. According to research, the personality trait that predicts this is if you tend to live in the moment rather than consider the results of your actions. This finding held true for both men and women—although men were more likely to be aggressive.[15]

THINK: WHY DO I WANT TO DRINK?

Which one (or more) of the following motives describes why you drink? It will help you decide which drinks are worth having or if another behavior might work instead.

1. You want to get smashed because it's exciting and different. You no longer want to be in control of your life.

2. Peer pressure. This might be buying rounds or drinking in a group. In a more extreme form, it's drinking games. In 2014, it was reported that up to five people died in the UK and Ireland doing the neck nomination or neknomination drinking game, an online dare to chug that you then had to pass on to another person. Thankfully this did not catch on in the US, where games like beer pong are more popular, but college drinking games are still associated with deaths from alcohol poisoning.[16] Of course, drinking games aren't new: the old-fashioned version was to down a glass of beer a yard tall.

3. You want an escape from having to think or feel, an emotional anesthetic that takes you out of the drudgery of life. After two to five drinks, as the endorphins and serotonin kick in, you no longer care so much about your problems, as your brain no longer functions that way.

4. For the laugh-out-louds and the highs. Dopamine and endorphins are what gives you the belly laughs and excitement so you don't know where the night will lead. You feel animated and interesting.

5. You want to have sex, and having a drink is how you feel brave enough to meet and/or talk to potential dates.

DRINKING AND DATING

There's a classic phrase, already mentioned, to describe how your perception of people distorts under the influence of alcohol: beer goggles. That's when a few drinks turn the man, woman, or person you don't fancy into someone who's extremely attractive.

In research that asked people about first dates, women said that drinking increased feelings of intimacy—defined as closeness, affection, and connectedness. For men, the opposite was true.[17]

In one study, people felt more positive toward pictures of others after a drink. But they didn't think art was better when they were drunk, which suggests it's a social effect rather than an aesthetic judgment.[18] And a second study showed that people find tipsy faces more attractive than sober ones—but drunk ones less so.[19]

So is drinking and dating a good thing? Not necessarily. In another study, 62 men in their 20s were sent on virtual dates with a computer simulation of a woman. Those who drank were much more persistent sexually in their conversations.

Probably in the majority of cases, alcohol lubricates what can be a tense and anxious situation, especially if it's a first date. At least there will be more relaxed conversation and increased good humor. And if one of you drinks more than is acceptable, then at least the other person has the option of turning down a second date.

HOW TO DRINK THE WAY YOU WANT TO (AND SENSIBLY)

SO YOU'VE DECIDED you want to drink less. As you know by now, I'm not a prohibitionist. I want you to get the maximum benefit out of drinking, with the least amount of harm.

The issue is, alcohol is more-ish. If you like to drink, it's always going to be easy to drink too much. In our alcohol-centered culture, if you say yes to every beer, glass, or bottle, you will end up drinking too much.

So it's up to you to discover the strategies that help you drink to the limit you decide, and that suit your finances, time limits, and lifestyle.

Ideally, if you find those that work, you'll be able to find an amount to drink that you are satisfied with, where you can function, have a good time, and feel comfortable with your level of health risk.

The first step is to think about what kind of drinker you are because the reasons you drink will make a big difference to the strategies that work. Next, I suggest a whole range of strategies, so you can pick and choose.

In this chapter, I'm not talking to dependent drinkers. If you are dependent, you will need medical attention during withdrawal and, most likely, a more structured approach to cutting down or giving up. Although some of these strategies may be part of that. For more, see Chapter 7.

WHAT KIND OF DRINKER ARE YOU?

You may think: Why do I need to know this? Without analyzing our drinking, we end up drinking too much. There are just too many opportunities to have alcohol. And, as I've said before, it's a strong drug. You wouldn't take medication or vitamins or an illegal drug without thinking about it first.

Most of the data over the last century has suggested there are two patterns of people who become alcoholic. These may not seem directly relevant to you, but they are enlightening about how alcohol can take hold of people.

The first group are young men with a family history of alcoholism. They become dependent almost from their first exposure to alcohol: with their first drink, everything makes sense.

The second group are people who drink to deal with the problems in their lives, whether those are anxiety, depression, or stress. They tend to turn to alcohol for self-medication from their 30s onward.

In the normal pattern of drinking over a person's lifetime, there are life phases when you are more vulnerable to drinking too much.

Young people start out drinking socially for pleasure, and often due to peer pressure too. They want to relax and they like the feeling of losing control.

Some people never stop drinking like this. But most give it up when life gets more responsible, whether that's due to a job or a relationship or a family.

Often, people will resume drinking later in life to deal with stress. And those with bigger issues, such as mental illness, grief, or PTSD, are particularly vulnerable in their search to suppress their symptoms.

One interesting way to classify your drinking comes from Professor Emmanuel Kuntsche and Dr. Sarah Callinan at the Centre for Alcohol Policy Research, La Trobe University, Australia.[1] It's useful because it helps you analyze your motivation to drink rather than all the specific one-off reasons you might have, such as a birthday or a bachelorette party. The idea behind this analysis is that your primary reason to drink is to change the way you feel, either to make more of positive feelings or to dampen down negative ones. Every other reason is secondary.

In this model, there are four types of drinkers. The categories are based on young adults, but in my experience, there are individuals of all ages who are typical of each of these categories, and most of us have some kind of overlap.

HOW TO DRINK THE WAY YOU WANT TO

1. **Social:** drinking mainly to celebrate and have fun.

   Are you preloading, drinking before you go out? Do you have a good time when you don't drink? Can you imagine a social occasion without drinking, or do you feel you need to have a drink in your hand? Social gatherings are difficult at any age, and social drinking can become a way to cope with them.

2. **Conformity:** drinking to fit in. You drink because other people do. If what your friends do is go out drinking, it's very difficult to be the one who doesn't drink.

   Are you drinking in rounds or doing shots, feeling you have to keep up with other drinkers? Does it feel rude to refuse a drink or someone topping up your glass? Would you be OK to be the person at the party who doesn't drink?

3. **Enhancement:** drinking because it's exciting. You might be a risk taker or do things when drunk that you wouldn't normally do. You're likely to go out with the intention of getting drunk.

   Do you crave the feeling of being drunk? Do you need to be drunk to let yourself go? Do you tend to binge drink? Don't wait until you have a bad experience or get injured or need a health intervention before you deal with your drinking behavior. If you do drink in order to get drunk, start to take cutting down very seriously.

4. **Coping:** drinking to forget your worries. This is the largest group, and in my experience includes a lot of middle-aged people who drink alone, self-medicating. These are the people who are most likely to become alcoholic. If you are drinking for stress reduction, do seek help rather than

papering over the cracks—or, rather, filling them up with vodka.

Do you drink every day? Do you use alcohol to numb your bad feelings? Are you drinking more than you used to? Are you able to spend a night alone without drinking? Start to think about ways you can learn to cope without alcohol.

PERSONAL HEALTH STRATEGIES FOR ALCOHOL: WHICH ONE WORKS?

First you need to really get a handle on your drinking: how much and when, as well as why and with whom.

WHAT IS A STANDARD DRINK IN THE US?

People who've never counted drinks are often shocked to discover how much they're drinking.

In the United States, one "standard" drink (or one alcoholic drink equivalent) contains roughly 14 grams of pure alcohol, which is found in

- 12 ounces of regular beer, which is usually about 5 percent alcohol
- 5 ounces of wine, which is typically about 12 percent alcohol
- 1.5 ounces of distilled spirits, which is about 40 percent alcohol

Of course the strength of beers, wines, and spirits vary quite a lot, so you have to do some simple math to allow for this. Also, when at home, you will be pouring your own drinks, so try to use the same size glass each time and use a measuring cup to see exactly the number of ounces you are pouring in each drink.

IS CUTTING DOWN RIGHT FOR ME?
OR SHOULD I GIVE UP?

If some of the following ring true for you, you might want to consider giving up rather than cutting down.

What follows are indications that cutting down might not be right for you:

1) You have tried and failed to give up before. You may want to seek help from your doctor because you may be dependent. The same is true if you drink heavily (see addiction advice, Chapter 7).

2) Your medical history. All of the following conditions/ habits/health risks have been linked to drinking and/or made worse by drinking. If any apply to you, you may be better off giving up.

- You have a mental health condition such as depression or anxiety.
- You suffer from gastric upsets: indigestion, gastritis, or reflux.
- You've given up smoking—or want to. (A lot of people relapse back to smoking when they drink, finding it to be a trigger.)

- You have high blood pressure.
- You have diabetes or prediabetes.
- You are at risk of stroke or heart attack, or you've had either.
- You have a family history of the cancers linked to alcohol (see pages 54–57).
- You have epilepsy. (You're more likely to have a fit when in withdrawal from alcohol.)
- You have a history of other drug use. This tends to go up when people drink.
- You want to conceive. This goes for both men and women.
- You have insomnia.
- You want to control your weight. Alcohol contains significant calories (see page 123).

3) Your personality. If your issue is that you lose control around alcohol, you may find it's better to make an active decision not to drink at all. It can be easier to say, "I don't drink," as opposed to, "I'm not drinking tonight," which can provoke the question, "Why, don't you like our company?" Peer pressure to drink can be quite powerful, and I often hear stories of people drinking when they didn't intend to.

If you go out intending to drink a little, you run the risk of failing. But if you decide not to drink and stick to that, people might think you're a little unusual but at least you've succeeded. And in fact, the good news is that being teetotal is becoming more socially acceptable, especially in younger age groups.

How to keep a drink diary

I've found keeping a daily drink diary to be the most useful way for people to get a handle on their drinking.

WEEK 1

• Fill out your drink diary

Include the time, where you were, who you were with, what you drank, how you were feeling and why. At the end of a week, for extra motivation, fill in a survey like the one at drinkmeter.com. It will compare your drinking with other people's and tell you whether or not you're drinking at dangerous levels.

• Write down why you're cutting down

Stick this up where you'll see it often, on the fridge, for example. Your reasons could be health related: I want to be healthier for my children, I want to have more energy, I want to feel more positive, I want to sleep better, I want to give up smoking. Or they could be linked to better performance: I want to run a 5k, I want to get a promotion at work. Or they could even be linked to looks: I want my skin to look more glowing, I want to lose weight, I want to age more slowly.

WEEK 2

• Eliminate the drinks that weren't worth having last week

Looking at your week of drinking, rank your drinks in order of fun, satisfaction, and results.

Which ones left you feeling good? Which could you have done without? Which do you wish you'd refused?

• Write down your triggers

On the nights you drank too much, what were you doing? Did you come straight out of work and, stressed, go to the bar to complain about it? Did you go home and pour yourself a large glass of red? Or was it that you were sitting alone and watching TV and felt like having a glass? Or that you were in the bar and everyone was buying rounds? Can you avoid these triggers? Work out what you'll do next time you're in that situation.

• Write down how you drink

Drink faster and you'll get drunk faster. Do you drink fast or slow? Does it depend on what you drink? Can you find a drink you can drink slowly? For example, you might find you drink red wine much more slowly than white. Can you slow down by alternating with soft drinks?

WEEK 3

• Set your target threshold

Decide how many standard drinks you're going to have next week. Calculate how many of your preferred drinks that is.

This has to be a threshold that you know you can keep to. If you set a threshold of a maximum of three drinks any one evening but you know you always lose control at four, then you should go lower. Remember, alcohol is one of the worst drugs for judgment.

● Be very specific

Look at your nights out and/or other possible drinking opportunities for the week and work out what you're going to drink on each occasion to avoid the "oh just one more" trap. Work out if you're going to drink beer or wine or spirits—then work out exactly what drinks you'll have. Think about how you are going to spread out your drinks over the week—bearing in mind you should have at least two drink-free days a week.

● Decide on a way to monitor your drinking

Some people use tokens to count drinks, swapping them from one pocket to another. Some people write it in a notebook or on a card they keep in their bag. Some people use apps, for example, Drinker's Helper. However, the truth is, if you're drinking sensibly you're very likely to have two drinks at the most, so you should be able to count them! However, many people do find it satisfying to write down their drinks.

If you go above your self-set limit, think about why this has happened to help you next week.

WEEK 4

● Change your routine

If you want to, which drinks can you cut out this week? Are there any that you're drinking just out of habit or that are no longer fun? Are you drinking out of stress? Adopt five of the suggestions in this chapter that make sense to you.

From now on, try out different strategies each week, until you reach the level of drinking that you like.

DRINK?

● Decide what your reward will be

Every time you keep within the limits you've set yourself, decide what your prize will be (tip: not a large glass of Merlot).

SENSIBLE DRINKING RULES YOU MIGHT WANT TO TRY

You don't need to do every one of these. In fact, there is some crossover between them, so you can't.

The idea is to try out a few you think might work and see if you want to—and can—adopt them permanently. I can't guarantee they will all work; some are evidence-based, some are personal rules that patients and colleagues have told me, and some I do myself.

What you are aiming for is a few new and lasting habits that help you keep to the limit you are happy with.

Don't drink when you're thirsty

Always satiate your thirst first with a soft drink or just water before you start on your usual full-strength drink. If you are a beer or lager drinker, you might even try a shandy. Some people find that if they drink beer first, they drink less over the evening because it fills them up—but bear in mind it's very easy to go over your limit when you are drinking beer, especially strong ales and lagers.

Never allow anyone to top up your glass

At parties and dinners where the wine is flowing, you're bound to be offered more before you've finished your glass. Make it a rule that you must empty the glass before it's refilled, so you can know exactly how many glasses you've had.

HOW TO DRINK THE WAY YOU WANT TO

Say no to champagne

The same goes for prosecco or any other sparkling wine. You're often offered it as your first drink on an empty stomach, the bubbles mean you get intoxicated faster, and at 12 percent ABV, it's relatively strong alcohol. This kind of drink might be part of social culture—but you can say no. Oh, and if you're offered champagne with dessert you can say no to that too. After all, it's likely you'll have drunk enough by that point.

Don't wash down food with wine

Have a jug of water and glasses on the table or a glass beside you when you're eating. Some people find they drink more sparkling water, or water with ice or a slice added; whatever makes it more appealing. Or you could make it a rule that you only drink wine between courses, not before the meal comes and not when the food is in front of you. Really savor the taste of each (small) mouthful.

Refuse aperitifs

There is an idea that having an aperitif makes the meal more pleasurable, that it's a necessary part of eating, but it's really just a way to get drunk fast.

Only have two drinks

Two drinks will make it very likely you'll stay within the recommended limits, whether you are having glasses of beer or 5-ounce glasses of wine.

Aim never to get drunk

If you feel drunk, it means you have probably consumed at least three or more standard drinks, maybe much more if you are a heavier drinker with more tolerance. And be aware that although you may not feel drunk compared with the people around you, whether that's true depends on who you're drinking with![2]

Never have one for the road

What an amazing concept "one for the road" is! You're ready to go home and you're likely to be tired. You've probably drunk enough.

Volunteer to drive

That way, you can legitimately refuse all alcohol without feeling you might offend. However, never drive anywhere you know you might be tempted to drink. Go by public transport if you can. It's far better to drink and be safe—even if you drink a little too much—than to drink and drive.

Take a week off drinking

This isn't as effective a reset as taking a month off (see "Dry January" on page 211). But it will help you reset your tolerance so you'll need less booze to get the same effect—more bang for your buck, essentially. One thing: Do not go straight back to drinking at the level you were before. Your tolerance will be lower, so you will get drunk faster. And if you have been a heavy drinker, it can be very dangerous.

Only buy booze as needed

Some people find it's best not to keep alcohol in the house. In any case, don't "build a cellar" to be there to tempt you. Instead, buy on the days you decide to drink.

Adopt a curfew

The traditional British naval saying "Never drink 'til the sun is over the yardarm" was supposed to stop people from drinking before noon! But you can create a healthier version with your own timings—for example, that you don't start drinking until 7 p.m. and you stop by 9:30 p.m.

Alternate your drinks

This is a great strategy if you're drinking, say, vodka and Coke: skip the vodka every other round. For beer and wine, drink a full glass of water between drinks. You can also make your drinks last longer by adding fizzy water or soda water to them.

Don't get alcohol delivered

Don't use a delivery service like PostMates or join a wine club, especially if it delivers you a selection automatically. They anonymize your relationship with the purchase. Instead, make yourself go to the shop so you can think about what you are choosing, why you are buying it, and when you're going to drink it.

Buy expensive alcohol

Or at least the most expensive you can. Thinking about how much money you're pouring down your throat may be a deterrent to drinking more than you want to. And you will hopefully also enjoy the flavor more.

Find low- and no-alcohol drinks you like

There are so many more on the market than there used to be—from alcohol-free spirits to soft drinks that aren't as sweet or childish. The bad news is, some of them are as expensive as alcohol, some even more.

More bars are now serving low-alcohol beer on tap, and if you want to hide the fact you're not drinking, a glass of this has to be the best way to do it. Most pubs now sell bottles of very low (0.5 percent) beer and some sell cider too. In reality they are practically nonalcoholic. For example, switching from normal Heineken to Heineken Zero will massively reduce your drinks total.

Change your patterns

Try one of the new ways to drink—or not drink. Some people call themselves 80–20 drinkers, so they drink two days out of ten. Or you could be a 5–2 drinker, only drinking two nights out of seven. Or a Weekend Warrior, who only drinks Friday, Saturday, and/or Sunday. Nota bene: don't binge; remember the data suggests bingeing is more dangerous in terms of brain effects and safety risks, among other health harms.

Find new activities

Sports leagues are the most obvious swap for the bars or drinking at home, as exercise and alcohol don't mix. Plus they're good if you're craving the friendly buzz of the bar. But any hobby or evening class will do. Been meaning to learn upholstery? Or do life drawing? Book it! On weekends, meet friends for walks or shopping or cups of tea instead of at a bar.

Book in drink-free days

The guidelines say two drink-free days a week is the minimum number to aim for. But even one is much better than none. Two consecutive nights will give your liver time to recover. As before, on the nights you do drink, don't binge.

Don't pregame

Because alcohol is expensive when you're out, you may be tempted to get tanked up before you go. But it's too easy to have all your drinks before you've even left the house.

Don't drink at home

Because alcohol is so much cheaper to buy from supermarkets and liquor stores than it ever has been, we are drinking more than ever at home. Some people even have special beer and wine fridges! But if the best part of drinking is being sociable, why not stick to only drinking when you're out?

Don't drink alone

Drinking alone is a habit you can do without; there's nobody to monitor you or make you rethink. And especially, don't drink in front of the TV or a screen, because when you do that, it becomes mindless. You can finish a bottle of wine before you've even realized.

DRINK?

If your state allows it, don't buy alcohol at the supermarket

Whether you buy as part of your regular shop or whether you add a few bottles to your supermarket delivery, stop. Instead, make alcohol a special shopping trip. You will probably buy less, and you will realize exactly how much you are spending too.

Avoid cheap alcohol

That means happy hours, which make you drink more in a shorter amount of time, and also supermarket offers. Bargains are difficult to resist, but if you take the alcohol home, you're unlikely not to drink it.

Make sure you're hydrated

Carry a refillable water bottle and keep drinking all day. The more hydrated you are, the less likely you are to down your first drink fast.

Enjoy what you drink

Choose alcohol you love the taste of, then savor it. Think of it as a delicious treat, not something to throw down your neck. Sip, don't gulp.

Never buy a wine box

They may seem like a good idea—after all, you won't feel you need to finish the bottle. But if you have one in your fridge or cupboard all the time, it makes drinking far too accessible.

Never open a second bottle

This is another one for the wine drinkers. If two of you share one 13 percent bottle of wine, you'll automatically limit yourself to 3.5 standard drinks each.

HOW TO DRINK THE WAY YOU WANT TO

Don't eat salty snacks

Salty snacks are put out in bars specifically to make you drink more. So avoid them at home too.

Treat your stress first

Plan a personal list of activities you can do to reduce your stress levels before you turn to drink—for example, exercise, meditation, yoga, breathing exercises, call a friend, have a bath, go for a walk.

Count before you go out

Say to yourself (or to your partner or a friend, to help you stick to it), "Tonight, I'm only going to drink two glasses of wine." And, if you know where you'll be at each stage of an evening out, work out exactly where and when you'll have those two glasses. So you might plan your night like this: orange juice and soda at the pub, one glass of wine when you arrive at the restaurant, one glass with the main course.

Practice saying "no, thank you"

Don't be afraid to say "no, thank you" to your host. It might help to practice; people can be very persistent. If you feel you need an excuse other than you're trying to drink less, think one up in advance. You can say you're driving, you're driving the morning after, you've got a big day the next day, you had a big night the previous night, you're on doctor's orders, you're on antibiotics or medication. Say whatever you feel comfortable with.

DRINK?

Order the smallest measure

If glasses were smaller, we'd drink less. A study from the University of Liverpool took place over four bar trivia nights. On two nights, the pub sold drinks in standard sizes: pints and wine in 175 milliliter glasses. On the other two nights, they sold beer in ⅔ pints and wine in 125 milliliter glasses. On the small-serving nights, people drank around a third less over the course of the evening.

It could be that people count their drinks, or it's too much effort to keep going back to the bar, or that smaller glasses make drinking a smaller amount seem the right thing to do.

Whatever the reason, it's worth ordering the smallest measure or glass when you're out. And investing in smaller glasses at home too. I have been known to use sherry glasses for dessert wine at home.

Turn down the music

A French study showed that loud music in bars makes you drink more quickly. It may be that that you become more excited and drink more as a result, or that you drink instead of talking because you can't hear.[3]

Don't get involved in rounds

Buy your own drinks; that way you will be in control. And you won't be tempted to drink to get your money's worth!

Use straight glasses

A study from the School of Experimental Psychology, University of Bristol showed that people drank lager 60 percent more slowly out of straight glasses than out of outward-curving (pilsner type) ones. The theory is, you pace yourself while drinking. And you drink a curved glass faster because it makes you think the halfway point is lower than it really is.

In fact the same group showed that if you put markings on the curved glasses, people controlled their drinking better then too. The takeaway? Knowing how much you're drinking helps you regulate your drinking.

Talk to your doctor

It's worth being honest with your doctor about how much you're drinking. They are not going to judge. The generally accepted rule of thumb among medics is that when you ask someone how much they drink, you then double their answer!

The doctor will use a brief questionnaire to work out how harmful your drinking is and whether you need specialist help or can cut down on your own (see page 148 for information on addiction). American research has shown that screening and one session of advice by your doctor can subsequently reduce the amount you drink at a single sitting by up to 25 percent.

Surround yourself with nondrinkers

There are new kinds of gatherings specifically for people who don't want to drink.[4] There are also nonalcohol bars being set up in the UK and also in the US, for example, Getaway in New York City.[5] Just like a carnivore might eat vegetarian food some nights, you don't have to be 100 percent teetotal.

Go out later

Research suggests that the later in the evening you start drinking, the less you will drink during the course of the night. You might expect that late starters would try to catch up, but they don't. In a study from the Netherlands, people who started between 8 p.m. and 9 p.m. drank far more over the course of an evening than those who started later. The late starters were also less likely to binge.

If you're used to being at the bar on the dot at 7 p.m., plan to go out at 8:30 or even 9 p.m. Sure, some of your friends might already be plastered. But that might even put you off drinking too much![6]

No shots or cocktails

Buying a round of shots is never a good idea. They can contain from one to two standard drinks in one hit. Cocktails can contain even more.

Half a drink an hour

When you're planning how much to drink, you could think of it as one in, one out. On average, you process around two thirds of a standard drink an hour. This strategy will take some sticking to, as it's only half a glass of wine or beer or cider per hour. But there is another plus: drinking at this pace will likely help you avoid a hangover too.

Get an app

If you're the kind of person who likes to track your progress, there are plenty of apps that do this. Try DrinkCoach or the Global Drug Survey Drinks Meter.[7]

HOW TO DRINK THE WAY YOU WANT TO

Don't drink at lunchtime

This is one of the easier rules to choose because it's getting increasingly unusual for lunchtime work functions to involve alcohol; I think because it's become clear that nobody does their best work when drunk, or even slightly inebriated. An alternative rule might be: never drink at work, including evening events.

Weddings and similar parties can mean starting drinking early in the day, sometimes with very little food. But if the party is likely to go on until the evening, why don't you start drinking when the evening guests arrive? Or after you've eaten? Think about how many more interesting conversations you'll have—that you'll actually remember.

Don't drink on an empty stomach

Line your stomach, ideally with some protein, fat, and whole-grain carbohydrate, and you'll slow down not only the rate at which you get drunk but also your peak blood-alcohol concentration levels—so you'll have less of a hangover too. Win-win!

Schedule an early morning event

Want an excuse not to drink? Book in a chore, appointment, or favor for the next morning for which you need to be awake, fully functioning, and sober. Ideally, make it something where other people are relying on you, so you're less likely to skip it, for example, giving someone a lift or meeting for a gym class.

Don't use energy drinks as mixers

A 2016 study showed that caffeine makes you drink more, probably by increasing the rewarding properties of alcohol.[8]

And a study the following year showed that energy drinks make you drink faster too.[9]

Buy the third round
If you only want to drink two glasses, buy the third round. That way you can order a nonalcoholic drink. You can even make sure it looks like alcohol so nobody can tell.

ESSENTIAL SAFER DRINKING RULES

Know your number
People need to know their alcohol intake in standard drinks the way they (should) know their calorie intake, their cholesterol level, and their blood pressure.

Make your intake target two standard drinks a day or fewer
This level of intake is not 100 percent "safe" but represents a threshold of harms that most authorities consider an acceptable level in relation to the pleasures of alcohol. Above these levels, the harms of alcohol begin to accelerate as intake increases.

Take pride in lowering your number
Just as lower numbers for blood pressure and cholesterol give greater life expectancy, so lower alcohol intake leads to improved longevity and health.

Take two drink-free days per week
The past 50 years have seen a major change in the use of alcohol in the US. Whereas in the 1950s, alcohol was seen as a

luxury, it is now widely perceived as a dietary staple and many people drink with every evening meal.[10] There is evidence that short periods of not drinking, even for just a day, can help the liver recover from the effects of alcohol and so reduce the risk of complications such as fatty liver and cirrhosis.[11]

WHY DRY JANUARY IS A GOOD IDEA

Taking a month off drinking has been shown to be a success for some people—and in real life, not just on social media.

It doesn't matter which month you choose—you could do Sober October or Dry July (although bear in mind that February is the shortest month!).

A survey by Dr. Richard de Visser at the University of Sussex questioned people who did 2018's Dry January. It found that even when it got to August of that year, the participants were drinking just 3.3 days a week, a drop from 4.3 days a week. Instead of getting drunk 3.4 times a month, they were down to 2.1 times a month. And they were drinking less on drinking days too: 5 standard drinks, down from their previous 6. Which is still way over the safe limits, but at least the trend is downward.

Of course, this is only a survey, so there's a risk that it's based on people's edited and possibly inaccurate views of their month. That said, probably the most long-term and life-changing benefits were in people's reported feelings about alcohol. Nearly eight out of ten said they understood when they felt more tempted to drink, and seven out of ten had learned they did not need alcohol to have fun. Other reported benefits? Better sleep, more energy, a sense of achievement, weight loss, and saving money.[12]

JUST SAY NO

If someone is used to you drinking, they might find it unsettling when you stop or cut down. So it can feel embarrassing and even insulting if you don't, for example, say yes to the wine at a dinner party, or to a round at the bar.

Because drinking is a social activity, peer pressure can be hard to resist. And that's not taking into account all the extreme drinking activities involved in joining some societies or how sports teams bond over drinking. Alcohol fuels that kind of group culture and vice versa.

I think drinking games—particularly hazing—should be banned. People die in drinking games. If someone wants you to take part in one, get away as fast as you can.

You shouldn't have to explain why you don't want a drink, but you probably will. People don't realize the harms of alcohol, so they still force drinks on others. But it's not really so different from forcing a cigarette on someone, and nobody does that, do they?

Another reason people may insist is that they prefer those around them to drink so they feel their level of drinking is normal.

Prepare an explanation as to why you're not drinking, or not as much as usual. You could say: "I don't want to get drunk because I don't want to embarrass myself." Or: "I am only drinking one glass because I believe that, for me, it's the right amount." Or even: "I feel good, I've reached the level of relaxation I was aiming for tonight."

If you are an alcoholic, you could simply say that. This would be a very powerful statement indeed.

A softer way to approach not drinking is: "I'm monitoring my drinking." After all, it's similar to monitoring calories and there's no shame in that. Or you could frame it in terms of calories: "I'm trying to lose weight and I've used up my calorie allowance."

If all else fails, and you're still being asked, follow the advice of this saying: "No is a complete sentence."

10

CHILDREN AND ALCOHOL

THE US SURGEON General's 2017 report (The 2017 Youth Risk Behavior Survey) found that among high school students, during the past 30 days: 30 percent drank some amount of alcohol, 14 percent binge drank, 6 percent drove after drinking alcohol, and 17 percent rode with a driver who had been drinking alcohol.[1] The advice from the UK's Chief Medical Officer is that children shouldn't drink at all before the age of 15. And that ages 15 to 17, they shouldn't drink more than once a week. That's because evidence shows that alcohol affects the development of bones and hormones, as well as the brain and other organs.

There are two factors that I think really matter at this early point in a person's life. The first is you don't want your child to be a statistic. Young people do die of alcohol poisoning. That means teaching them how to be safe if they are going to drink now. The second is about the long term. If your child does decide to drink, you want them to learn a sensible way of drinking that will mean they don't harm their health in the long term.

It's true that some teens don't use alcohol at all now, which must be a relief to parents. But that's not true of all teens: WHO's *Global Status Report on Alcohol and Health 2018* showed that in Europe, 44 percent of 15- to 19-year-olds do drink.

If your teen wants to drink, you are unlikely to be able to stop them from having any alcohol. More worryingly, nearly one in five 15-year-olds think it's OK to get drunk once a week.

People often ask me if they should let their child drink at home, or at a party. That is a hard question. If you have or have had an alcohol problem yourself, I'd say the answer is probably no. They may have a similar genetic or other predisposition to alcoholism as you.

Otherwise, introducing alcohol—in the form of beer or cider of a strength less than 4 percent, or very diluted wine—on special occasions at home from the age of 15 is probably fine, although it's a myth that it will stop your child binge drinking. Bear in mind that statistics show the earlier your child drinks, the more likely they will have problems with alcohol later in life (see Fig. 6 on page 216). Of course this doesn't prove that drinking alcohol earlier leads to dependence, as there may be genetic or environmental factors, but it is worth knowing.

Even if you don't have an alcohol problem, think about what your own drinking is saying to your child: Is your habit of finishing a bottle of wine a night giving the opposite message to your lecture on moderation? Has your child seen you drinking to get drunk? Of course children will do what you do, not what you say. And don't be the person who buys them drinks.

I think the secret to helping your child drink sensibly—as well as drinking sensibly yourself—is education. Tell them how to use alcohol safely.

Explain that people can and do die from alcohol poisoning after just one night's drinking. And the reason why we don't always hear about it in the news is that it is common enough not to be news.

Fig. 6. Prevalence of lifetime alcohol dependence according to age of drinking onset

Source:
https://www.niaaa.nih.gov/news-events/news-releases/age-drinking-onset-predicts
-future-alcohol-abuse-and-dependence.

You can explain that getting drunk is more likely to lead to fights, accidents, injuries, unwanted sexual experiences, and being mugged, as well as doing things they will regret such as a university student in the UK who was filmed peeing on the wreaths of a war memorial and was lucky not to be sent to prison.[2]

HOW TO HELP TEENS DRINK SAFELY AT A PARTY

If your teen is having a party at home, the key is supervise, supervise, supervise. If your child is going to a party at another child's house, can you speak to the parents and make sure they are going to be there?

Bear in mind that this advice is for age 15 plus. Before that, I'd advise no alcohol at all at parties. Also, be aware that in some US states, there can be serious legal ramifications for parents caught hosting underage drinking.

- Don't go out. Be obviously present. Do not trust teenage children to the care of older children—they won't take any notice of them.
- Confiscate any strong alcohol and spirits as the teens come in.
- Confiscate any drugs. My message about drinking and drugs is very simple: don't drink and drug.
- If any child looks drunk, take away their alcohol, put them somewhere quiet where they can sit down, and give them some water. Make sure someone is looking after them and that they get home safely.
- If any child passes out and can't be roused, put them in the recovery position and call an ambulance.
- Confiscate all car keys and only release them to someone who is clearly sober.

HOW TO TALK TO YOUR CHILDREN ABOUT BOOZE

Alcohol should be a sensible conversation throughout your child's life, not just when they ask to have their first party with alcohol. Obviously what you say needs to be made age-appropriate: the following list is aimed at a 15-year-old.

1) **IT WON'T MAKE YOU HAPPY** Alcohol affects your mental health—it can lead to anxiety and depression, or make them worse. A recent UK study, *Smoking, Drinking and Drug Use among Young People 2018*, showed that a third of young people ages 11 to 15 who'd recently drunk alcohol had low levels of happiness. For those who hadn't recently drunk, it was 22 percent.[3]

2) **AVOID SPIRITS** You won't have a sense of how to regulate your drinking yet. So it's way better to drink weak beer than strong spirits, in particular, shots. With spirits, you can get catastrophically drunk before you've even realized it. In fact, it's a good idea to avoid any alcohol that is more than 4 percent ABV.

3) **YOU CAN DIE FROM ALCOHOL POISONING** According to the WHO, alcohol is the leading cause of death in young people—especially in wealthy countries like the US.[4]

4) **BE AWARE YOU'RE BEING SOLD ALCOHOL** Drink advertising is clever at seducing you into wanting to drink, by associating drinking with sex, money, and fun. The alcohol industry wants you to start drinking. And research suggests that the more alcohol advertising teenagers see, the more they drink. Your favorite sport is likely to have an alcohol brand as a sponsor—Anheuser-Busch sponsors the NFL, for example. And the same

might be true of the last event or concert you went to. The hospitality industry—aka the alcohol industry— also sponsors sporting and cultural events and invites influential people, from celebrities to politicians to journalists.

5) HAVE ONE DRINK (OR TWO MAX) If you have one drink, you'll have fun, but you shouldn't get drunk (if it's 4 percent beer or cider). If you have two, you may be drunk.

6) IF YOUR FRIEND IS DRUNK, HELP THEM Don't do what the friend of then-UK Prime Minister Tony Blair's son Euan is said to have done—assume he'd gone home when, in fact, he was lying unconscious in London's Leicester Square.[5] Make sure you get your friend home. Don't be frightened about telling their parents—they will very likely be grateful to you. They'd rather their child was safe.

Your friend may not just be asleep. If they are unconscious and unresponsive, call an ambulance. Don't do what many so-called friends do and make fun of them or do horrible things to them.

7) DRUNK PEOPLE GET INTO TROUBLE They have accidents and fights, commit criminal acts, get arrested. One doctor I know, when she was a medical student, woke up in a stranger's house. The man explained she had knocked on his door the night before, so drunk she couldn't speak. Having no idea who she was, he had let her sleep on his sofa. She'd gone home to the wrong house. She couldn't remember any of this. She was lucky she'd stayed safe.

8) GET HOME SAFELY You are vulnerable after drinking because you are less aware of, for example, a mugger coming up behind you and less likely to recognize or care about danger. Stay with your group of friends. Don't go home alone. Don't get into a stranger's car. Don't get an unlicensed taxi. Don't walk home if it's not safe or you wouldn't do it sober. Work out how you'll get home before you go out.

9) DON'T DRINK AND DRUG Never, ever take drugs when you are drinking. One, you can't predict the result of mixing the two. And two, you may not know what you're taking or you won't be aware enough to be certain.

In 2010, two boys in Eastern England, Louis Wainwright, 18, and Nicholas Smith, 19, were reported to have died of mephedrone, aka meow meow, an Israeli drug that was originally created for controlling aphids.

CNN called me for a comment on the case; I said I was very surprised at this because in an Israeli review of 400,000 people who'd taken this stimulant, there had been no deaths.[6]

Later, the real story came out. The boys had been drinking all evening in several bars, then when the bars shut had gone into the town to buy drugs. It turned out that instead of mephedrone, they had bought and taken the heroin substitute methadone. It might have been that they were so drunk they didn't know what they took. Also, if you take drugs when drunk, you're more likely to die. Methadone and all other opioid drugs suppress breathing and alcohol accentuates that effect.

10) DON'T DRIVE WHEN DRINKING This should go without saying. In the US over 10,000 people die every year from drunk-driving accidents.[7] Other things not to do if you've had a drink: climbing and/or jumping off high things, especially into the sea from a pier or into a river or reservoir or pool. In fact, don't go swimming. And never give or take dares.

11) WEAR CLOTHES As alcohol dilates (widens) your blood vessels, it makes you feel warm even when it's cold outside. So even in winter, people who've been drinking will go out without a coat. But when you are drinking, you also lose heat faster, so it's easier to freeze to death—especially if you sit down and then fall asleep outside. So wear proper clothes. And don't ever stop and sit down on the way home—especially not in a snowdrift.

12) LEARN FROM YOUR MISTAKES If your friends tell you what an idiot you were last night, believe them and learn from it. You may not remember—be thankful that someone does and is watching out for you.

13) DON'T POST OR MESSAGE Or email or text or go on social media. You will likely regret it.

14) WALK AWAY FROM FIGHTS People can get incredibly violent when drunk and really hurt one another. They get disinhibited and easy to offend. I have seen a fight where people were glassed. I hid under a table, watching the blood pour out of them. If a fight starts, walk—or run—away.

15) KNOW WHAT MAKES FOR CONSENT It is not possible to give consent while you are drunk. So don't have sex with anyone who is drunk (or who is under the age of consent).

One in three women have been sexually taken advantage of while under the influence of drugs or alcohol, mostly in private homes by someone they know, according to the 2019 *Global Drug Survey*. Nearly 90 percent of these cases involved alcohol.

CONCLUSION

OVER THE PAST 50 years—my scientific lifetime—there have been massive advances in the science of alcohol. We now have a much clearer picture of how its toxic properties affect almost all areas of the body. Whereas before we just knew the link between alcohol and liver cirrhosis, we now also know the particular contribution it makes to high blood pressure, with its knock-on effects of heart attacks and strokes. And we know alcohol use is associated with at least eight cancers too. The risk of cancer from drinking is not as high as the risk of lung cancer from smoking but it is still significant.

Another area of major advance in knowledge is the actions of alcohol on the brain. When I started in medicine in the 1970s it was believed alcohol was a solvent, that it dissolved into the cell walls of neurons and so disrupted their activity. We now know this is a minor action and that what's really important is how alcohol acts on many different neurotransmitter systems. For example, alcohol enhances the activity of the sedating inhibitory neurotransmitter GABA but also blocks the activity of the wakefulness-promoting neurotransmitter glutamate. This double whammy is why severe alcohol intoxication can be so dangerous and why it can lead to coma and even death.

As well as affecting GABA and glutamate, alcohol produces a cocktail of effects on other neurotransmitters. It increases the stimulant neurotransmitter dopamine, the pleasure peptide endorphin, and also the empathy-producing neurotransmitter serotonin. It's this special cocktail that makes alcohol so pleasurable, especially socially. But it's also its effect on these neurotransmitters that can lead to dependence and addiction, especially in people with genetic predispositions or those who are using alcohol to deal with stress or other mental health issues.

This book also shares with you the challenges that governments and policy-makers face in dealing with the harms alcohol produces combined with our desire to drink and the influence of the immensely rich and powerful alcohol industry. I believe there are proven policies that rational governments could instigate to maximize the benefits and minimize the harms of alcohol in our society. I have campaigned all my professional life for these. The challenge is getting the media and you—the general public—to understand this and campaign for them too. I hope that this book gives you the information and motivation to support me in this ambition.

US ALCOHOL DRINKING RECOMMENDATIONS

THESE ARE BASED on the guidance given by the Surgeon General, the Centers for Disease Control, and the National Institute of Alcohol Abuse and Alcoholism. They state that no level of alcohol consumption is safe but recommend that if you do drink, then try to comply with the moderate drinking guidelines such as those produced as part of the Dietary Guidelines for Americans. In these, moderate drinking is defined as up to 1 standard drink per day for women and up to 2 drinks per day for men.

They further point out that binge drinking is the most common form of excessive drinking, and this is defined as consuming
- For women, 4 or more drinks during a single occasion
- For men, 5 or more drinks during a single occasion

Heavy drinking is defined as consuming
- For women, 8 or more drinks per week
- For men, 15 or more drinks per week

Importantly, they also point out that most people who drink excessively are not alcoholics or alcohol dependent, which puts the imperative to consider our drinking levels onto the majority of the population.

Source: https://www.cdc.gov/alcohol/fact-sheets.htm.

RESOURCES

Alcoholics Anonymous: https://www.aa.org

Al-anon (for family members of alcoholics): https://al-anon.org

National Association for Children of Alcoholics (NACOA): (888) 554-COAS, https://nacoa.org

National Council on Alcoholism and Drug Dependence, Inc. (NCADD): 1-800-NCA-CALL (622-2255)

National Institute on Alcohol Abuse and Alcoholism Alcohol Policy Information System: https://alcoholpolicy .niaaa.nih.gov

Samaritans HELPLINE: (877) 870-HOPE (4673)

SAMHSA's National Helpline: 1-800-662-HELP (4357)

NOTES

PREFACE

1 https://www.washingtonpost.com/news/wonk/wp/2015/10/16
/the-hidden-cost-of-your-drinking-habit/.

INTRODUCTION

1 www.bmj.com/content/362/bmj.k3944.
2 www.bmj.com/bmj/section-pdf/986178?path=/bmj/363/8172/This
_Week.full.pdf.
3 https://www.washingtonpost.com/news/wonk/wp/2014/09/25
/think-you-drink-a-lot-this-chart-will-tell-you/.
4 https://www.washingtonpost.com/lifestyle/2019/06/24/when-teens
-drink-its-different-than-adults-heres-how-you-can-help-your
-child-avoid-problem/.
5 en.wikipedia.org/wiki/Weihenstephan_Abbey.
6 www.bmj.com/content/362/bmj.k3944.
7 www.thelancet.com/journals/lanpub/article/PIIS0140-6736(10)61462
-6/fulltext;www.ncbi.nlm.nih.gov/pubmed/25922421;journals.sagepub
.com/doi/abs/10.1177/0269881119841569?journal-Code=jopa.
8 doi.org/10.1002/ijc.27553.

HOW DRINKING AFFECTS YOUR BODY AND BRAIN

1 www.ncbi.nlm.nih.gov/pubmed/28216062.
2 Additional information from: www.alcohol.org/effects/blood -alcohol-concentration.
3 pdfs.semanticscholar.org/9bb9/846bf3eec9fd7b1a37ae6e0e8cc0 25d34563.pdf.
4 foodanddrink.scotsman.com/drink/the-controversial-story-of -buckfasts-rise-to-prominence-in-scotland.
5 www.ncbi.nlm.nih.gov/pmc/articles/PMC2426682.
6 www.ncbi.nlm.nih.gov/pubmed/27380261.
7 www.ncbi.nlm.nih.gov/pmc/articles/PMC543875.
8 theconversation.com/lining-your-stomach-with-milk-before-a-big -night-out-and-other-alcohol-myths-88116.
9 pubs.niaaa.nih.gov/publications/aa72/aa72.htm.
10 www.researchgate.net/publication/263287740 _Editorial_Can_Hangover_Immunity_be_Really_Claimed.
11 www.researchgate.net/publication/26693901 _Are_Some_Drinkers_Resistant_to_Hangover_A_Literature _Review.
12 www.cdc.gov/media/releases/2015/p1015-excessive-alcohol.html.
13 www.medicalnewstoday.com/articles/325619.php.
14 www.ncbi.nlm.nih.gov/pmc/articles/PMC4230485.
15 www.ncbi.nlm.nih.gov/pubmed/22434663.
16 www.sciencedaily.com/releases/2018/05/180514095530.htm.
17 news.bbc.co.uk/1/hi/health/8416431.stm.
18 www.ncbi.nlm.nih.gov/pmc/articles/PMC5515685.
19 journals.sagepub.com/doi/pdf/10.1177/2050324517741038.

THE HEALTH HARMS OF ALCOHOL

1 www.researchgate.net/publication/235971838_Alcohol_consump tion_alcohol_dependence_and_attributable_burden_of_disease _in_Europe_Potential_gains_from_effective_interventions_for _alcohol_dependence.

NOTES

2   https://www.niaaa.nih.gov/publications/brochures-and-fact-sheets/alcohol-facts-and-statistics.

3   www.sciencedaily.com/releases/2018/10/181003102732.htm.

4   understandinguncertainty.org/files/2012bmj-microlives.pdf.

5   www.ncbi.nlm.nih.gov/pubmed/27073140.

6   https://www.cdc.gov/alcohol/faqs.htm.

7   www.ncbi.nlm.nih.gov/pubmed/15770105.

8   www.dailymail.co.uk/news/article-5843059/NHS-pharmacy-worker-37-died-toxic-poisoning-misjudging-alcohol-measures-holiday.html.

9   www.nytimes.com/2018/07/18/health/cirrhosis-liver-cancer.html.

10 www.nhs.uk/conditions/alcohol-related-liver-disease-arld/treatment.

11 https://www.ucsf.edu/news/2019/01/413096/liver-transplants-double-alcohol-related-liver-disease.

12 Personal communication from Professor Nick Sheron.

13 eprints.soton.ac.uk/427505/1/1_s2.0_S246826671830241X_main.pdf.

14 www.ncbi.nlm.nih.gov/pmc/articles/PMC3992057.

15 www.drinkaware.co.uk/alcohol-facts/health-effects-of-alcohol/diseases/alcohol-and-liver-cancer.

16 www.wcrf.org/dietandcancer/exposures/alcoholic-drinks.

17 www.aicr.org/continuous-update-project/reports/breast-cancer-report-2017.pdf.

18 www.wcrf.org/dietandcancer/exposures/alcoholic-drinks.

19 cordis.europa.eu/project/rcn/92876/reporting/en.

20 https://theconversation.com/how-alcohol-damages-stem-cells-and-increases-cancer-risk-new-research-89578.

21 www.medicalnewstoday.com/articles/314539.php.

22 www.thelancet.com/journals/lancet/article/PIIS0140-6736(18)31772-0/fulltext.

23 www.vox.com/science-and-health/2018/8/29/17790118/alcohol-lancet-health-study.

24 Milwood et al., 2019 Lancet on Kadoorie study.

25 www.ahajournals.org/doi/full/10.1161/01.HYP.33.2.653.

26 www.ahajournals.org/doi/pdf/10.1161/strokeaha.108.520817.

27 www.thelancet.com/journals/lanpub/article/PIIS2468-2667(17)30003-8/fulltext.

28 www.bmj.com/content/357/bmj.j2353.

29 www.bmj.com/company/newsroom/both-long-term-abstinence-and-heavy-drinking-may-increase-dementia-risk.

30 www.ahajournals.org/doi/10.1161/CIRCULATIONAHA.108.787440.
31 www.nytimes.com/2018/06/18/health/nih-alcohol-study.html.
32 www.ncbi.nlm.nih.gov/pubmed/26294775.
33 www.ncbi.nlm.nih.gov/pubmed/29103170.
34 www.ncbi.nlm.nih.gov/pmc/articles/PMC4590619.

AAA: ALCOHOL, ACCIDENTS, AND AGGRESSION

1 https://pubs.niaaa.nih.gov/publications/surveillance106/HDS14
.htm.
2 https://www.cdc.gov/vitalsigns/crash-injuries/index.html.
3 www.gov.iruk/government/statistics/reported-road-casualties-in
-great-britain-final-estimates-involving-illegal-alcohol-levels
-2017.
4 https://www.responsibility.org/media_center/media_centermajor
ity-of-americans-do-not-know-legal-limit-for-drunk-driving
-signs-of-alcohol-impairment/.
5 www.ncbi.nlm.nih.gov/pmc/articles/PMC4814299.
6 www.alcoholhelpcenter.net/Program/BAC_Standalone.aspx.
7 www.researchgate.net/publication/223136111_Alcohol-Related
_Risk_of_Driver_Fatalities_An_Update_Using_2007_Data.
8 Ibid.
9 webarchive.nationalarchives.gov.uk/20100921035247/http:/northre
view.independent.gov.uk/docs/NorthReview-Report.pdf.
10 www.sussex.ac.uk/broadcast/read/48912.
11 etsc.eu/wp-content/uploads/report_reducing_drink_driving_final.pdf.
12 www.ias.org.uk/uploads/pdf/bloodalcoholcontenteffectivenessre
view.pdf.
13 eprints.gla.ac.uk/189646.
14 https://www.nytimes.com/2019/12/23/business/drunk-driving
-interlock-crash.html.
15 www.oisevi.org/a/archivos/estudios-especificos/ong/Union
-Europea-Druid-Final-Report.pdf.
16 www.ncbi.nlm.nih.gov/pubmed/18955613.
17 www.bath.ac.uk/announcements/why-the-effects-of-a-boozy-bank
-holiday-binge-could-last-longer-than-you-think.

18 www.ncbi.nlm.nih.gov/pubmed/20497803.
19 en.wikipedia.org/wiki/The_Gulag_Archipelago.
20 www.historyextra.com/period/viking/the-truth-about-viking -berserkers.
21 www.sciencedaily.com/releases/2009/02/090218081624.htm.
22 www.inverse.com/article/49228-who-alcohol-report-drinking -deaths.
23 https://www.bjs.gov/content/acf/29_prisoners_and_alcoholuse .cfm.
24 www.bbc.co.uk/news/uk-england-20630771.
25 metro.co.uk/2011/06/17/ascot-fight-on-ladies-day-stuns-horse -racing-fans-47597/.
26 alcoholchange.org.uk/publication/are-you-looking-at-me.
27 theconversation.com/if-england-gets-beaten-so-will-she-the-link -between-world-cup-and-violence-explained-99769.
28 www.ias.org.uk/uploads/IAS%20report%20Alcohol%20domes tic%20abuse%20and%20sexual%20assault.pdf.

WHAT ALCOHOL DOES TO YOUR MENTAL HEALTH

1 https://www.ncbi.nlm.nih.gov/pmc/articles/PMC5397789/.
2 www.nhs.uk/news/mental-health/alcohol-and-depression.
3 www.cmaj.ca/content/191/27/E753.
4 pubs.niaaa.nih.gov/publications/arh26-2/130-135.htm.
5 adaa.org/understanding-anxiety/social-anxiety-disorder/social -anxiety-and-alcohol-abuse.
6 www.cambridge.org/core/journals/the-british-journal-of-psychia try/article/loss-of-consciousness-and-posttraumatic-stress-disor der/E82CD29771CBD412EBA3D51A896CB059.
7 www.ncbi.nlm.nih.gov/pmc/articles/PMC3770804.
8 www.ncbi.nlm.nih.gov/pubmed/29183241.
9 academic.oup.com/qjmed/article/99/1/57/1523792.
10 www.verywellmind.com/adult-alcoholism-adhd-connected-63078.
11 www.sciencedirect.com/science/article/abs/pii/088761859290009V.
12 jamanetwork.com/journals/jamapsychiatry/article-abstract /497548.

13 www.sciencedirect.com/science/article/abs/pii/S03064
60398000094.
14 www.ncbi.nlm.nih.gov/pubmed/6598815.
15 www.sciencedirect.com/science/article/abs/pii/S0306460396000536.
16 link.springer.com/article/10.1007/s00127-005-0981-3.
17 www.ncbi.nlm.nih.gov/pubmed/22983943.

HORMONES AND FERTILITY

1 www.ncbi.nlm.nih.gov/pmc/articles/PMC1772851.
2 www.ncbi.nlm.nih.gov/pmc/articles/PMC3527168.
3 www.sciencemediacentre.org/expert-reaction-to-research-on
-alcohol-intake-and-male-fertility.
4 www.bionews.org.uk/page_137273.
5 www.sciencedaily.com/releases/2014/10/141002221232.htm.
6 Ibid.
7 www.bmj.com/content/354/bmj.I4262.
8 www.nature.com/articles/s41598-017-14261-8.
9 www.ncbi.nlm.nih.gov/pubmed/3367299.
10 https://www.usnews.com/news/health-news/articles/2019-04-25
/drinking-among-pregnant-women-increases-in-us.
11 pubs.niaaa.nih.gov › publications › arh22-1.
12 www.sciencedaily.com/releases/2018/11/181128154000.htm.
13 www.nhs.uk/conditions/foetal-alcohol-syndrome.
14 bmjopen.bmj.com/content/8/12/e022578.
15 www.ncbi.nlm.nih.gov/pmc/articles/PMC5609722.
16 www.ncbi.nlm.nih.gov/pmc/articles/PMC4445685.
17 en.wikipedia.org/wiki/The_South_African_Wine_Initiative#cite
_note-AFP-2.
18 www.theguardian.com/global-development/2018/may/27/mothers
-children-foetal-alcohol-syndrome-south-africa.
19 www.sciencedaily.com/releases/2017/08/170821122756.htm.
20 Ibid.
21 academic.oup.com/humupd/article/22/4/516/2573866.
22 www.ncbi.nlm.nih.gov/pubmed/16500807.
23 www.breastcancer.org/research-news/20080311.
24 www.sciencedaily.com/releases/2017/11/171107092906.htm.

25 www.ncbi.nlm.nih.gov/pubmed/24118767.
26 https://www.cdc.gov/breastfeeding/breastfeeding-special
-circumstances/vaccinations-medications-drugs/alcohol.html.
27 www.sciencedirect.com/science/article/abs/pii/S0306453011003350.
28 www.pnas.org/content/pnas/early/2009/06/26/0812809106.full.pdf.
29 www.ncbi.nlm.nih.gov/pubmed/30738971.

HOW ALCOHOL AFFECTS
YOUR QUALITY OF LIFE

1 www.nhs.uk/apps-library/sleepio.
2 science.howstuffworks.com/life/sleep-obesity1.htm.
3 www.sciencedaily.com/releases/2017/04/170420114020.htm.
4 jech.bmj.com/content/71/12/1177.
5 www.sussex.ac.uk/broadcast/read/47131.
6 www.independent.co.uk/life-style/health-and-families/health-news
/the-incredible-shrinking-man-5544531.html.
7 www.euro.who.int/—data/assets/pdf_file/0018/319122/Public-health
-successes-and-missed-opportunities-alcohol-mortality-19902014
.pdf.
8 www.researchgate.net/publication/26314858_Beer_consumption
_and_the_"beer_belly"_Scientific_basis_or_common_belief.
9 www.thelancet.com/pdfs/journals/lanpub/PIIS2468-2667(17)30089
-0.pdf.
10 https://www.niznikhealth.com/research-articles/drinking-in-the
-workplace.
11 https://www.theatlantic.com/business/archive/2013/07/the
-economic-cost-of-hangovers/277546/.
12 www.sciencedaily.com/releases/2018/08/180825124245.htm.
13 medicalxpress.com/news/2019-08-hangovers-brain-function.html.
14 https://www.ncbi.nlm.nih.gov/pmc/articles/PMC6518120/.
15 www.drinkaware.co.uk/alcohol-facts/health-effects-of-alcohol/life
style/can-alcohol-affect-sports-performance-and-fitness-levels.
16 www.ncbi.nlm.nih.gov/pmc/articles/PMC4629692.
17 https://www.northpointrecovery.com/blog/back-deep-end
-michael-phelps-went-alcohol-rehab-greatest-olympian-ever.

18 https://www.thrillist.com/drink/nation/baseball-drinking-stories
-mickey-mantle-babe-ruth-harry-caray-drink-beer-booze-in-the
-major-leagues.
19 https://www.thefix.com/brett-favre-opens-about-rehab-stays
-alcoholism-addiction.
20 www.theguardian.com/science/sifting-the-evidence/2015/nov/17
/sex-does-alcohol-really-make-you-better-in-bed.
21 www.ncbi.nlm.nih.gov/pmc/articles/PMC1403295.
22 www.sciencedaily.com/releases/2016/08/160804141034.htm.
23 www.sciencedaily.com/releases/2016/09/160918214439.htm.
24 www.sciencedirect.com/science/article/pii/S205011611830120X.
25 www.ncbi.nlm.nih.gov/pmc/articles/PMC2917074.
26 www.sciencedaily.com/releases/2014/03/140328102907.htm.
27 www.forbes.com/sites/chunkamui/2016/03/22/wine-and-sleep
-make-for-better-decisions/#38f815a824b1.
28 www.sciencedirect.com/science/article/pii/S1053810016303713.
29 www.forbes.com/sites/chunkamui/2016/03/22/wine-and-sleep
-make-for-better-decisions/#38f815a824b1.
30 en.wikipedia.org/wiki/Rum_ration.
31 https://en.wikipedia.org/wiki/Beer_day.
32 www.telegraph.co.uk/sport/othersports/darts/6637884/Andy
-Fordham-I-became-world-darts-champion-despite-never-being
-sober.html.
33 www.dailymail.co.uk/news/article-2794310/drunk-belgian
-anaesthetist-killed-british-mother-caesarean-told-police-need
-vodka-don-t-shake.html.
34 www.theguardian.com/world/2014/oct/03/british-woman-dies
-botched-caesarean-france.
35 www.bmj.com/content/2/5103/993.
36 DOI: 10.1177/0269881117735687.

ADDICTION: HAVE I GOT AN
ALCOHOL PROBLEM?

1 http:// www.washingtonpost.com/news/wonk/wp/2014/09/25/think
-you-drink-a-lot-this-chart-will-tell-you/.

2 pdfs.semanticscholar.org/60cd/d49b5ed6a84423762a366727ae4880 c6f255.pdf.

3 https://jamanetwork.com/journals/jamapsychiatry/fullarticle /2647079.

4 psycnet.apa.org/record/1977-27694-001.

5 americanaddictioncenters.org/alcoholism-treatment/symptoms -and-signs/hereditary-or-genetic.

6 www.ncbi.nlm.nih.gov/pubmed/28734091.

7 www.ncbi.nlm.nih.gov/pubmed/8611056.

8 Ibid.

9 www.ncbi.nlm.nih.gov/pubmed/15704214.

10 www.alcohol.org/alcoholism/is-it-inherited.

11 en.wikipedia.org/wiki/Disulfiram-like_drug.

12 americanaddictioncenters.org/alcoholism-treatment/symptoms -and-signs/hereditary-or-genetic.

13 www.imperial.ac.uk/news/192468/scientists-untangle-links -between-genes-intake.

14 www.jsad.com/doi/pdf/10.15288/jsa.1977.38.1004.

15 www.cochrane.org/CD004148/ADDICTN_effectiveness-brief -alcohol-interventions-primary-care-populations.

16 www.ncbi.nlm.nih.gov/pubmed/11912067.

17 www.niaaa.nih.gov/news-events/news-releases/niaaa-reports -project-match-main-findings.

18 ajp.psychiatryonline.org/doi/full/10.1176/appi.ajp.2016.16050589.

19 https://www.datafiles.samhsa.gov/study/national-survey-drug-use -and-health-nsduh-2018-nid18757.

20 www.independent.co.uk/news/uk/home-news/student-death-initi ation-ceremony-newcastle-university-ed-farmer-drunk-a8602061 .html.

21 www.ncbi.nlm.nih.gov/pmc/articles/PMC3286419.

THE SOCIAL BENEFITS OF ALCOHOL

1 www.sirc.org/publik/drinking3.html.

2 medovina.com/history.htm.

3 Roueche, B., "Alcohol in Human Culture," in S. Lucia, ed., *Alcohol and Civilization*. New York: McGraw-Hill, 1963, p. 171.

4 www.ncbi.nlm.nih.gov/pubmed/21954999.
5 medium.com/pebble-research/the-secret-to-happiness
   -2348011c8497#.q95rbpl1m.
6 link.springer.com/article/10.1007/s40750-016-0058-4.
7 eprints.qut.edu.au/83729/4/Ryan_McAndrew_Thesis.pdf.
8 www.ncbi.nlm.nih.gov/pubmed/3981386.
9 www.sciencedaily.com/releases/2017/05/170515091131.htm.
10 www.psychologicalscience.org/news/releases/moderate-doses-of
   -alcohol-increase-social-bonding-in-groups.html.
11 news.bbc.co.uk/1/hi/health/3035442.stm.
12 www.kent.ac.uk/news/society/9730/alcohol-can-make-you
   -momentarily-happier.
13 www.sciencedirect.com/science/article/abs/pii/S0277953
   616301344.
14 www.sciencedaily.com/releases/2017/05/170515091131.htm.
15 www.sciencedaily.com/releases/2011/12/111219135215.htm.
16 https://web.archive.org/web/20080711133153/http://www.cnn
   .com/2008/HEALTH/07/07/drinking.game.deaths.ap/index
   .html?iref=mpstoryview.
17 www.sciencedaily.com/releases/2018/07/180705114017.htm.
18 www.ncbi.nlm.nih.gov/pubmed/30717614.
19 www.sciencedaily.com/releases/2018/07/180705114017.htm.

HOW TO DRINK THE WAY YOU
WANT TO (AND SENSIBLY)

1 drinktank.org.au/2018/03/there-are-four-types-of-drinker-which
   -one-are-you.
2 graziadaily.co.uk/life/food-and-drink/heres-reason-sometimes-get
   -drunk-quickly.
3 www.sciencedaily.com/releases/2008/07/080718180723.htm.
4 https://www.npr.org/sections/health-shots/2019/06/23/732876026
   /breaking-the-booze-habit-even-briefly-has-its-benefits.
5 https://nypost.com/2019/04/23/nycs-sober-bar-scene-is-a-hip
   -oasis-for-booze-free-fun/.
6 www.researchgate.net/publication/328678730_Young_Adults_Do

_Not_Catch_Up_Missed_Drinks_When_Starting_Later_at_Night
-An_Ecological_Momentary_Assessment_Study.

7 https://www.globaldrugsurvey.com/brand/drinks-meter/.
8 www.ncbi.nlm.nih.gov/pmc/articles/PMC5344720.
9 www.ncbi.nlm.nih.gov/pubmed/27419377.
10 https://www.ncbi.nlm.nih.gov/books/NBK217463/.
11 www.ncbi.nlm.nih.gov/pubmed/17344205.
12 www.publications.parliament.uk/pa/cm201012/cmselect/cmsctech
/1536/153602.htm.
13 www.sussex.ac.uk/broadcast/read/47131.

CHILDREN AND ALCOHOL

1 https://www.cdc.gov/alcohol/fact-sheets/underage-drinking.htm.
2 www.mirror.co.uk/news/uk-news/drunken-student-charged-after
-urinating-425067.
3 digital.nhs.uk/news-and-events/latest-news/pupils-who-have
-recently-smoked-drunk-alcohol-and-taken-drugs-are-more-likely
-to-be-unhappy-according-to-new-report.
4 http://www.who.int/mediacentre/factsheets/fs349/en/#.
5 www.dailymail.co.uk/news/article-1309143/Euan-I-young-pints
-drunk-I-lost-Family-friend-night-Blairs-son-arrested.html.
6 Nutt, D. J., *Drugs: Without the Hot Air*. Cambridge: UIT Press,
2012.
7 https://www.cdc.gov/motorvehiclesafety/impaired_driving
/impaired-drv_factsheet.html.

ACKNOWLEDGMENTS

Most important thanks are to Brigid Moss for her great research and writing skills and to Sam Raim for editing the US version. Also, to Rachel Mills for finally working out what was the best book for me to extract from my research and clinical career.

Also to my many friends and colleagues who have taught me about alcohol over many decades but especially Christer Allgulander, Ilana Crome, Dora Duka, Ian Gilmore, David Goldman, Tony Goldstone, Paul Glue, George Koob, Anne Lingford-Hughes, Markku Linnoila, Richard Lister, Jurgen Rhem, Marcus Munafo, Charles O'Brien, Robin Room, Nick Sheron, Julia Sinclair, and Wim van den Brink. And for breakthrough insights in harm assessment and communication, Larry Philips and David Spiegelhalter.

And my wife, Di, who has for 40 years put up with my chaotic academic and clinical life; and our daughter Suzy, who runs Abbotshill, our wine bar in Ealing, and has taught me the value of quality organic wine!

ABOUT THE AUTHOR

DAVID NUTT

MBBChir (Cambridge) DM (Oxford), FRCP, FRCPsych,
FMedSci, hon DLaws (Bath)

David Nutt is a psychiatrist and the Edmund J. Safra Professor
of Neuropsychopharmacology in the Division of Brain Science,
Department of Medicine, Hammersmith Hospital, Imperial
College London. His research area is psychopharmacology—
the study of the effects of drugs on the brain, from the perspectives of both how drug treatments in psychiatry and neurology
work, and why people use and become addicted to some
drugs, such as alcohol. To study the effects of drugs in the
brain, he uses state-of-the-art techniques, such as brain imaging with PET and fMRI plus EEG and MEG. This research
output has led to over 500 original research papers and puts
him in the top 0.1 percent of researchers in the world. He has
also published a similar number of reviews and book chapters, eight government reports on drugs, and 35 books, including one for the general public, *Drugs: Without the Hot Air*,
that won the Transmission Prize in 2014.

He was previously president of the European Brain Council,
the British Association of Psychopharmacology, the British

Neuroscience Association, and the European College of Neuropsychopharmacology. He is currently founding chair of DrugScience.org.uk, a charity that researches and tells the truth about all drugs, legal and illegal, free from political or other interference. He also currently holds visiting professorships at the Open University and University of Maastricht.

David broadcasts widely to the general public on both radio and television. In 2010, the *Times Eureka* science magazine voted him one of the 100 most important figures in British Science, and he was the only psychiatrist on the list. In 2013, he was awarded the John Maddox Prize from Nature/Sense about Science for standing up for science and in 2017 a Doctor of Laws honoris causa from the University of Bath.

en.wikipedia.org/wiki/David_Nutt
www.sciencemag.org/content/343/6170/478.full
www1.imperial.ac.uk/departmentofmedicine/divisions/brainsciences/psychopharmacology